HIGH SCHOOL MATH MADE UNDERSTANDABLE BOOK 1: MATH 9 & 10

WRITTEN BY: JEREMY MARTIN

Copyright 2021 by: Jeremy Martin

Published by: Jeremy Martin

ISBN: 9798662712744

Revision: 2

All rights reserved. This book may not be reproduced in whole or in part without written permission from the publisher, except by a reviewer who may quote brief passages in a review; nor may any part of this book be reproduced, stored in a retrieval system, or transmitted in any form or by any means, electronic, mechanical, photocopying, recording, or other, without written permission from the publisher.

All the graphs in this book were created by using *DESMOS* web graphing application. Written permission was acquired from DESMOS prior to using it for the creation of Unit 3 in this book. Each graph has an attribution to *DESMOS* at the bottom-left corner.

ACKNOWLEDGEMENTS

I would like to thank my mother (Carole Martin), my father (Daniel Martin), and my sister (Angelique Martin), for their support throughout High School and throughout the creation of this book. I would also like to thank Cordell Wiebe, the Math teacher at GROW Centre(the school where I graduated), in Williams Lake, BC, who helped me with Math whenever I got stuck or stumped with a new concept.

Table of Contents

INTRODUCTION 8

POLYNOMIALS (INTRODUCTION) 12

EXPONENT LAWS...24

LINEAR FUNCTIONS 30

TRIGONOMETRY (Introduction)....................................52

FACTORING POLYNOMIALS 68

VOLUME & SURFACE AREA OF 82

GEOMETRIC FIGURES...82

RADICALS 92

Bonus Chapter: Trigonometry (11)..............................104

Answers & Solutions to Practice Questions..................124

INTRODUCTION

Thank you for purchasing *High School Math Made Understandable Book 1*, I hope you enjoy reading this book just as much as I enjoyed writing it. I wrote this book with the aspiration of making High School Math understandable for everyone. Many mathematics books assume you already have a solid understanding of some algebra concepts and will jump right into some new complex topics, leaving many readers confused if they do not have a solid comprehension of basic algebra. In this book I will assume you have no background knowledge in Algebra, and we will begin with the most basic subjects in Algebra, Geometry, Trigonometry, and Radicals. Each chapter is divided into several sections. At the end of each chapter there are practice questions that you should attempt before moving on to the next chapter. Once you have completed the questions then you can check the answers in the *Answers & Solutions to Practice Questions* section that is found near the end of this book. The solutions section of this book gives you the step by step solution to the practice questions, this way if you get the wrong answer you can see where you went wrong. I strongly suggest you attempt the problem before looking at the answer and solution, by attempting the problem by yourself you will learn much more than just by copying answers from the back of the book. Now before we start, there is one more thing I want to tell you, motivation, dedication, and persistence are the three most important factors that play a role in your learning of math.

For additional Math Learning resources make sure to check out **Jerematics.Weebly.com** where you can find the link to my YouTube Channel (Jerematics Tutoring)!

Formula Sheet

Trigonometry

$$\sin\theta = \frac{opp}{hyp} \qquad \cos\theta = \frac{adj}{hyp} \qquad \tan\theta = \frac{opp}{adj}$$

Pythagorean Theorem: $c^2 = a^2 + b^2$

Linear Equations

Slope: $\dfrac{rise}{run} = \dfrac{Y_2 - Y_1}{X_2 - X_1}$ Slope-intercept form: $y = mx + b$

Order of Operations

BEDMAS : **B**rackets **E**xponents **D**ivision **M**ultiplication **A**ddition **S**ubtraction

POLYNOMIALS (INTRODUCTION)
Chapter 1

What is a **polynomial**? A *polynomial* is an expression consisting of **variables** (a quantity which can change, represented by a letter) and **coefficients**, that involves the operations of addition, subtraction, multiplication, and non-negative integer exponents of variables. Non-negative integer exponents mean that the exponent is below 0, or even a variable with an exponent that is a fraction.

Here are a few examples of what a polynomial is compared to what is not considered a polynomial:

Examples of Polynomials	Not Polynomials
$x + 2$	4^{-2} (negative exponent)
$x^2 + x + 2$	
$a^2 + b + 1$	$\sqrt{x}$ (Variable is inside radical)
5	
	$\frac{1}{x}$ (Variable is in denominator)

Here are the three basic *terms*:

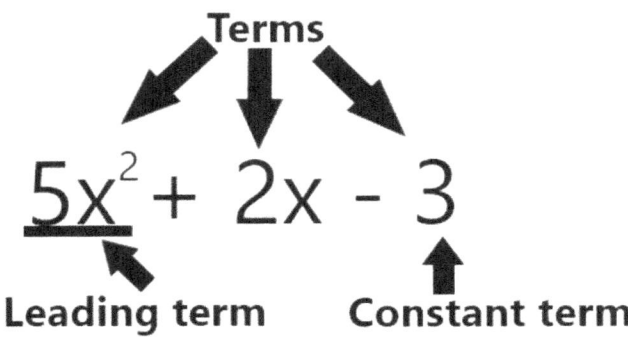

The *middle term* is simply called the "*Middle* term". The order of the terms is determined by the degree of the variable. The *degree* is the exponent of the variable, the larger the exponent the greater the degree is and the variable with the highest degree (exponent) goes first. An exponent may also be called a *Power*.

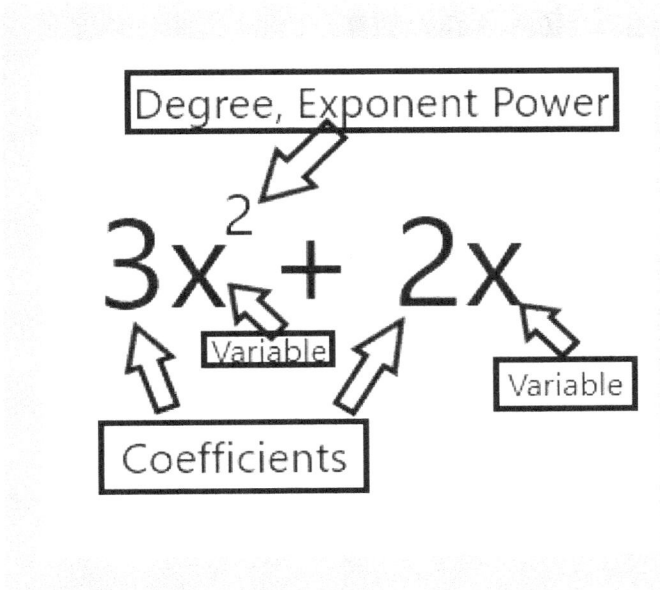

One of the first skills you will learn will be to collect "**like terms**" to simplify an expression. A like term is determined by the variable portion of the term, they must contain the **same variable(s)** and **exact same power(s)** to be able to be combined. Once you have determined that two terms are *"like terms"* they can be combined.

Here are a few examples of "like" and "unlike" terms, they have been placed inside brackets, so it is easier to read:

- (**x**) and (**x**) are "like" terms.
- $(3x^2y^2)$ and $(2x^2y^2)$ are "like" terms (both variables and exponents are the exact same, coefficient in front does not matter).
- (**x**) and (x^2) are **not** "like" terms.
- (**x**) *and* (**xy**) *are* **not** "like" terms.

Adding Like Terms

When adding "like terms" together, you add the coefficients of the two like terms together like this:

$3x + 5x = 8x$ since, 3+5 = 8

Here are some more examples:

Example 1: $x + x = 1x + 1x = 2x$

Example 2: $2x + 3x = 5x$

Example 3: $3y + 7y = 10y$

Example 4: $2x^2 + 3x$ cannot be simplified further (they are not like terms)

Example 5: $2x^2 + 4x^2 = 6x^2$

If there are no coefficients in front of the variable, the coefficient is assumed to be 1, there is no use in putting a 1 in front of the variable that is why there was no coefficient in front of the two *x*'s in the first example.

The best way to think about collecting like terms (the x's, y's, b's, etc.) is to think of it as a fruit. For example, if in a bowl of fruit, you have an apple, an orange, another apple, and another orange, how many apples and oranges are there? Well there are two apples and two oranges, you add one apple with the other apple to get two apples, you do not count it as an orange because it is an apple, not an orange. The same thing goes for "like terms" the "x" goes with the "x" not the "y." The "y" goes with the "y."

Each different variable can be seen as its own unique "Object," for instance, "ab" is its own unique object and can be added to another variable that is exactly the same, which would be another "ab".

Subtracting like terms

When you are subtracting "like terms," you subtract the coefficients of the terms like this:

$7y^2 - 2y^2 = 5y^2$ since,   7-2 =5

Here are some more examples:

Example 1: $x - 2x$

$= 1x - 2x$

$= -1x$

$= -x$

Example 2: $x^3 - x$ cannot be simplified (they are not like terms).

Example 3: $5y - 2y = 3y$

Example 4: $7xy^2 - 3xy^2 = 4xy^2$

Evaluating Expressions With A Given Value For Variables

Sometimes you will be asked to evaluate an expression with a given value for a variable, what you do is you basically just replace the variable with the given value, multiply the coefficient of the term with the value of the variable, and then simplify the expression.

For example,

$x = 3, y = 2$

$x + y + 1 = ?$

Step 1. Substitute "x" with **3**. Substitute "y" with **2**.

(3) + (2) + 1 =?

Step 2. Simplify,

3 + 2 + 1 = 6

In the expression $x + y + 1$

When x = 3 and y = 2, the sum of the expression is 6.

Here are two more examples:

$x = 2,\ y = 3$	$x = a,\ y = b$
Expression: $5x - 2y = ?$	Expression: $3x + 2y = ?$
$= 5(2) - 2(3) = 10 - 6$	$= 3(a) + 2(b)$
$= 4$	$= 3a + 2b$

As you can see the variable(s) is(are) replaced with a value or another letter.

Solving Algebraic Expressions

To solve for a variable in an equation, you must "isolate" the variable on one side of the equation and write the equation in terms of that variable.

Remember to eliminate something from one side of the equation, you must perform the opposite operation on both sides of the equals sign. Read this table from left to right:

Case	Solution	Example
Addition on one side (+)	Subtract on both sides (-)	Example: $x + 2 = 5$ subtract 2 from both sides $x + 2 - 2 = 5 - 2$ $x = 3$
Subtraction on one side (-)	Addition on both sides (+)	Example: $x - 2 = 7$ add 2 to both sides $x - 2 + 2 = 7 + 2$ $x = 9$
Multiplication on one side (•)	Division on both sides (÷)	Example: $3x = 9$ Divide both sides by 3

		$\frac{3x}{3} = \frac{9}{3}$ $x = 3$
Division on one side ($\div$)	Multiplication on both sides ($\cdot$)	Example: $\frac{x}{4} = 4$ Multiply both sides by 4 (to cancel out the denominator) $\frac{x}{4}(4) = 4(4)$ $x = 16$
Squaring one side (x^2)	Square root both sides ($\sqrt{}$)	Example: $x^2 = 9$ Square root both sides $\sqrt{(x^2)} = \sqrt{9}$ $x = \pm 3$ $\pm$ means (plus or minus), two answers because a negative number multiplied by itself will always result in a positive number
Square root one side ($\sqrt{}$)	Square both sides side (x^2)	(This is an algebraic expression not a polynomial expression) Example: $\sqrt{x} = 5$ $(\sqrt{x})^2 = (5)^2$

		$x = 5(5)$ $x = 25$

Order of Operations (BEDMAS)

In this section we will look at a very important topic, *Order of Operations*. The order of operations is a certain order you must follow to correctly perform all mathematical operations. The order you should follow is:

Brackets : anything inside of the → ( )

Exponents : Anything such as → 2^3, $((x+5)^2)$, etc

Division : when you have a division statement such as $3 \div 2$

Please note that $/$ and $\overline{}$ also denotes division

Multiplication : (*) or (·) A number or variable multiplied by another

Addition: (+)

Subtraction: (-)

We will look at a few examples (multiplication and division of variables will be covered in the next chapter).

Example 1: Simplify $(2+5) - 2 * 3$

Start with the brackets → $(7) - 2 * 3$

Multiplication $-2 * 3$ → $(7) - (2 * 3) = 7 - 6$

Subtract → $7 - 6 = 1$

Answer is 1

Example 2: Simplify $\dfrac{6}{2} + (3 - 1) + x + x$

Brackets first → $\dfrac{6}{2} + (2) + x + x$

Division → $3 + 2 + x + x$

Collect like terms → $= 5 + 2x$

Answer is $5 + 2x$

Practice Questions

1. Simplify, $2x + 2x + 3x + y$

2. True or False:
   a) x and x^2 are like terms?
   b) $3x^2$ and $2x^2$ are like terms?
   c) x and y are like terms?

3. Simplify the following expressions:
   a) $3y - y$
   b) $2x^2 + 3x^2$
   c) $ab + ab$

4. Evaluate the following expressions if $x = 2$, and $y = 4$:
   a) $2x + 3y$
   b) $y - x$
   c) x^2
   d) y^2

5. Solve the following algebraic expressions for x.
   a) $3x = 4 - x$

b) $2x + 2 = 8$

EXPONENT LAWS
Chapter 2

An **exponent** is basically a shorthand for a number repeatedly multiplied by itself. The exponent indicates how many times a number or variable should be multiplied by itself. The **base** number tells what number is being multiplied. The *exponent*, the small number written above and to the right of the base number, tells how many times the *base* number is being multiplied.

Here is an example:

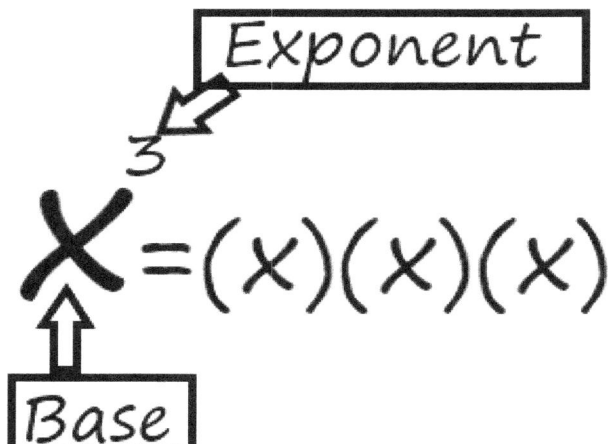

As you can see, the exponent determines how many times the base gets multiplied by itself. Another general rule is that any base to the power of 0 is 1. Ex: $x^0 = 1$. Now that you know the difference between a base and an exponent, we will now look at some "**Laws of Exponents**" that will be useful when simplifying expressions.

Law 1: *law of multiplication of identical bases.*

When multiplying two or more identical bases, you simply add the two exponents together.

Law 1: $x^a * x^b = x^{a+b}$

Here are some examples,

Example 1: $x^3 * x^2 = x^{3+2} = x^5$

Example 2: $x * x = x^1 * x^1 = x^{1+1} = x^2$

Example 3:
$2^3 * 2^3 = 2^{3+3} = 2^6 = (2)(2)(2)(2)(2)(2) = 64$

Example 4: $(x^2)(y^3)$ *we cannot simplify example 4 any further as they are not identical bases.*

Example 5: $x^3 * x^2 y^2$ we can simplify the "x" base but not the other base "y"

$$x^3 * x^2y^2 = (x^3)(x^2)(y^2) = (x^{3+2})(y^2) = x^5y^2$$

Now that we have looked at the basics of the first law, we will now look at the next law, the law of dividing identical bases.

Law 2: *law of Dividing of identical bases,*

When dividing two identical bases, you simply subtract the exponent of the base in the denominator from the exponent of the base in the numerator.

Law 2: $\dfrac{x^a}{x^b} = x^{a-b}$

Here are some examples:

Ex 1: $\dfrac{x^3}{x^2} = x^{3-2} = x^1 = x$

Ex 2: $\dfrac{x^a y^c}{x^b y^d} = x^{a-b} y^{c-d}$

Ex 3: $\dfrac{x^5 y^6}{x^2 y^3} = x^{5-2} y^{6-3} = x^3 y^3$

Ex 4: $\dfrac{4^6}{4^3} = 4^{6-3} = 4^3 = (4)(4)(4) = 64$

Law 3: *Power raised to a Power,*

If you have a product inside parentheses, and a power on the parentheses, such as [$(x^2)^3$], then you simply multiply the inner exponent by the outer exponent.

$$(x^a)^b = x^{a*b}$$

Word of Warning: This rule does NOT work if you have a sum or difference within parentheses. Exponents unlike multiplication do NOT distribute over subtraction or addition. Ex: $(x^2 + y^2)^2$ Is **Not** equal to $(x^4 y^4)$, $(x^2 + y^2)^2$ is **Actually** equal to $(x^2 + y^2)(x^2 + y^2) = x^4 + y^4 + 2x^2 y^2$ we will cover this in greater detail later on in this book.

Here are some examples of when this third law can be applied:

Ex 1: $(x^2)^3 = x^{2*3} = x^6$

Ex 2: $(x^2y^3)^4 = x^{2*4}y^{3*4} = x^8y^{12}$

Ex 3:
$(2^3)^2 = (2^{3*2}) = 2^6 = (2)(2)(2)(2)(2)(2) = 64$

Ex 4: $(x^3)^y = (x^{3*y}) = x^{3y}$

Practice Questions

1. Simplify this expression: $y(xy^2)$, *Hint: it involves the first law of exponents.*

2. Simplify, $\dfrac{a^3c^2}{a^2}$

3. Simplify, $(y^2)^a$

4. Simplify, $\dfrac{x^2y^2}{x^2y^2}$

5. Simplify, $\dfrac{x^y}{x^z}$

6. Simplify and evaluate, $\dfrac{7^5}{7^2}$

LINEAR FUNCTIONS
Chapter 3

Before we begin learning about **Linear Functions**, we will do a brief review on the Cartesian Plane. Points on the plane are in terms of *x* and *y*, these are the coordinates. *x* is the **Horizontal** coordinate, meanwhile, *y* is the **Vertical** coordinate. Coordinate are written like this: (x, y), as you can see *x* is first and *y* is second.

Example 1: Let us plot the point (2,3), The black dot represents the point,

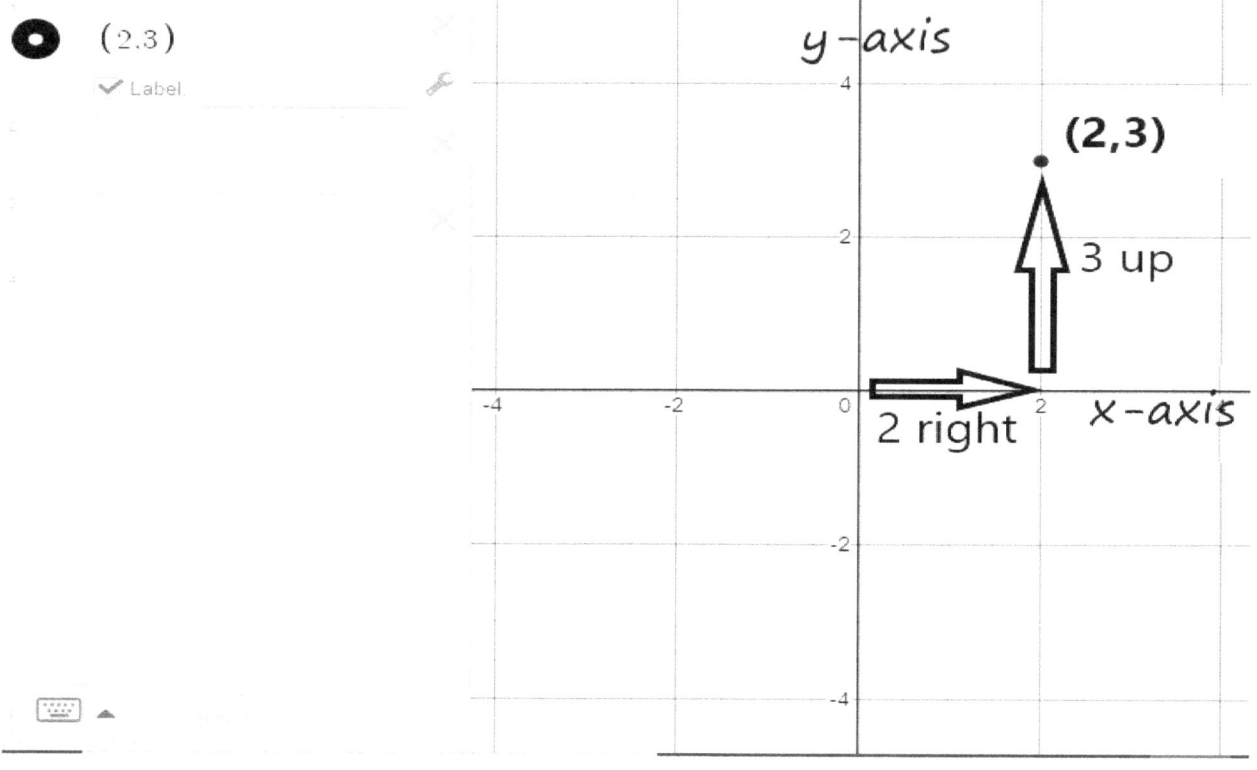

Example 2: Point (-4,-2), means 4 units left and 2 units down,

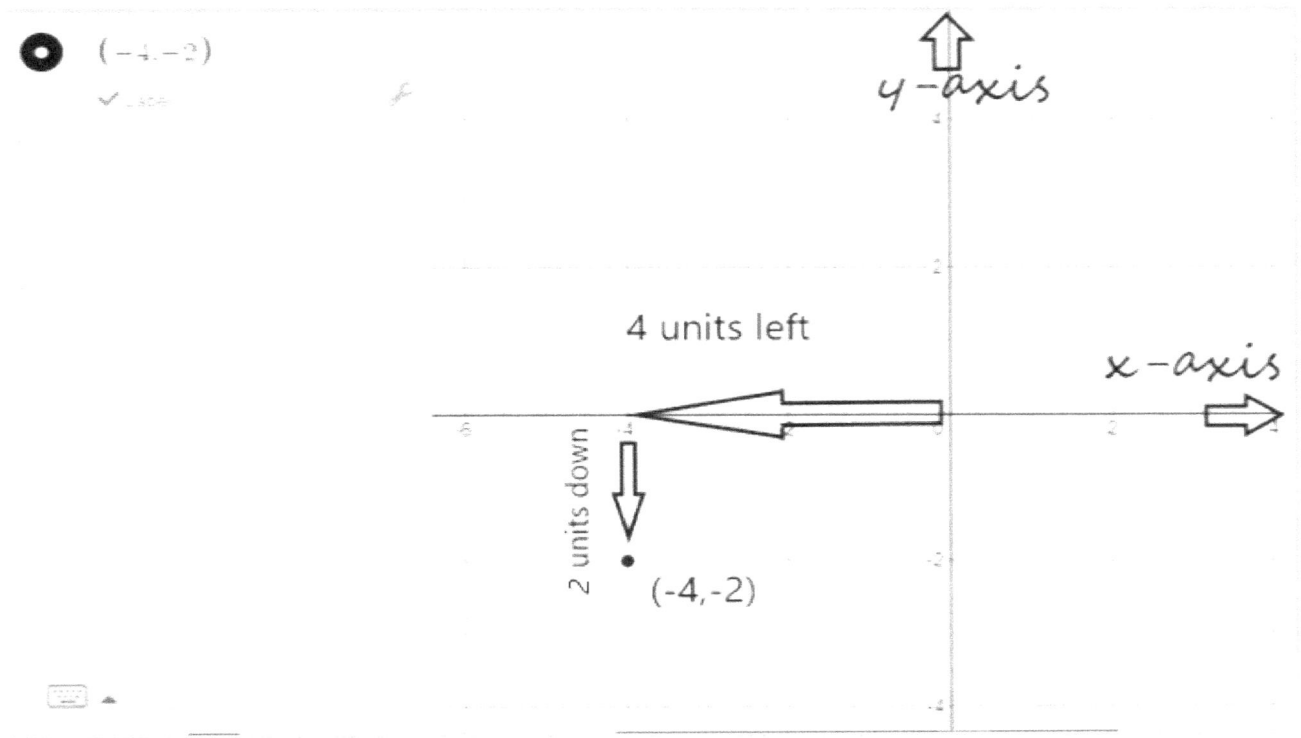

An *x-coordinate* above zero (+) means "number of units right," an *x-coordinate* below zero (-) means "number of units left." A *y-coordinate* above zero (+) means "number of units up," a *y-coordinate* below zero (-) means "number of units down."

The Cartesian Plane is divided into 4 quadrants, I (1), II (2), III (3), IV (4). Point (0,0) has a special name, it is called the **Origin**.

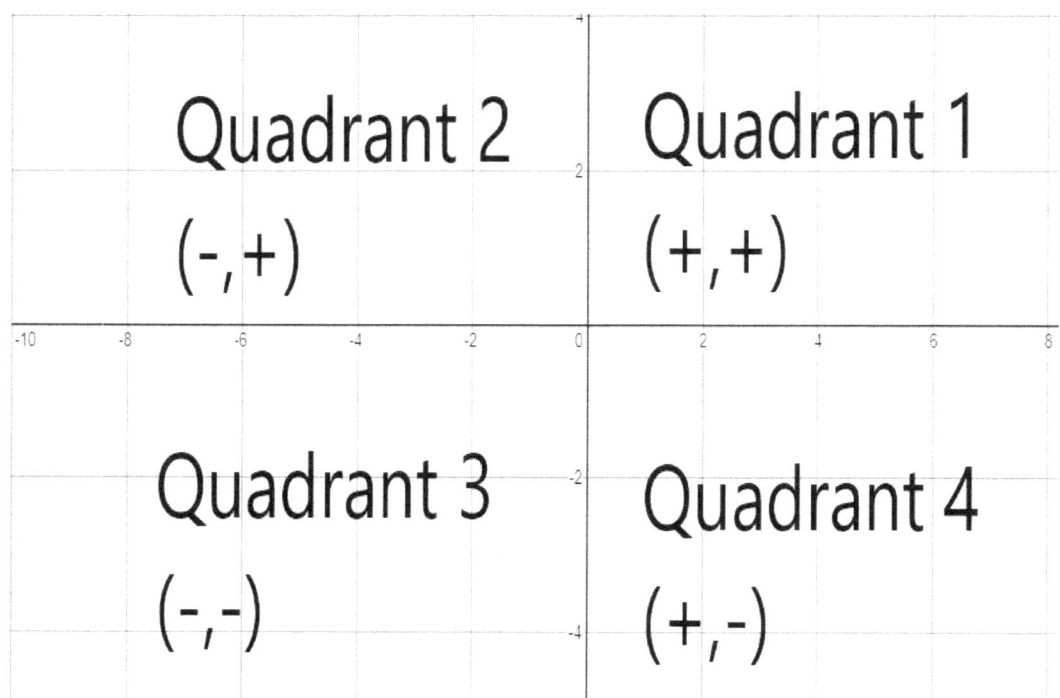

Here are some features of each Quadrant:

Quadrant 1: x and y coordinates are both positive.

Quadrant 2: x coordinate is negative, and y coordinate is positive.

Quadrant 3: x and y coordinates are both negative.

Quadrant 4: x coordinate is positive and y coordinate is negative.

All Linear Functions covered in this chapter are continuous meaning that it goes to positive and negative infinity on both axis. The most basic linear function is:

$y = x$

Y is the dependent variable and x is the independent variable, this means that the y-value depends on the value of x, it relies on the input. X is considered your input value, and Y is considered the output value. Think about it this way, input for example is what you enter on your calculator, what is displayed on your screen depends on what you enter on your calculator. The output is what is displayed on your screen on the calculator when you press the "=" button. This is the result, just like in the equation $y = x$ the final value that results, depends on what you put inside of it (x-value). Here is the graph of the most basic equation you will see in High School:

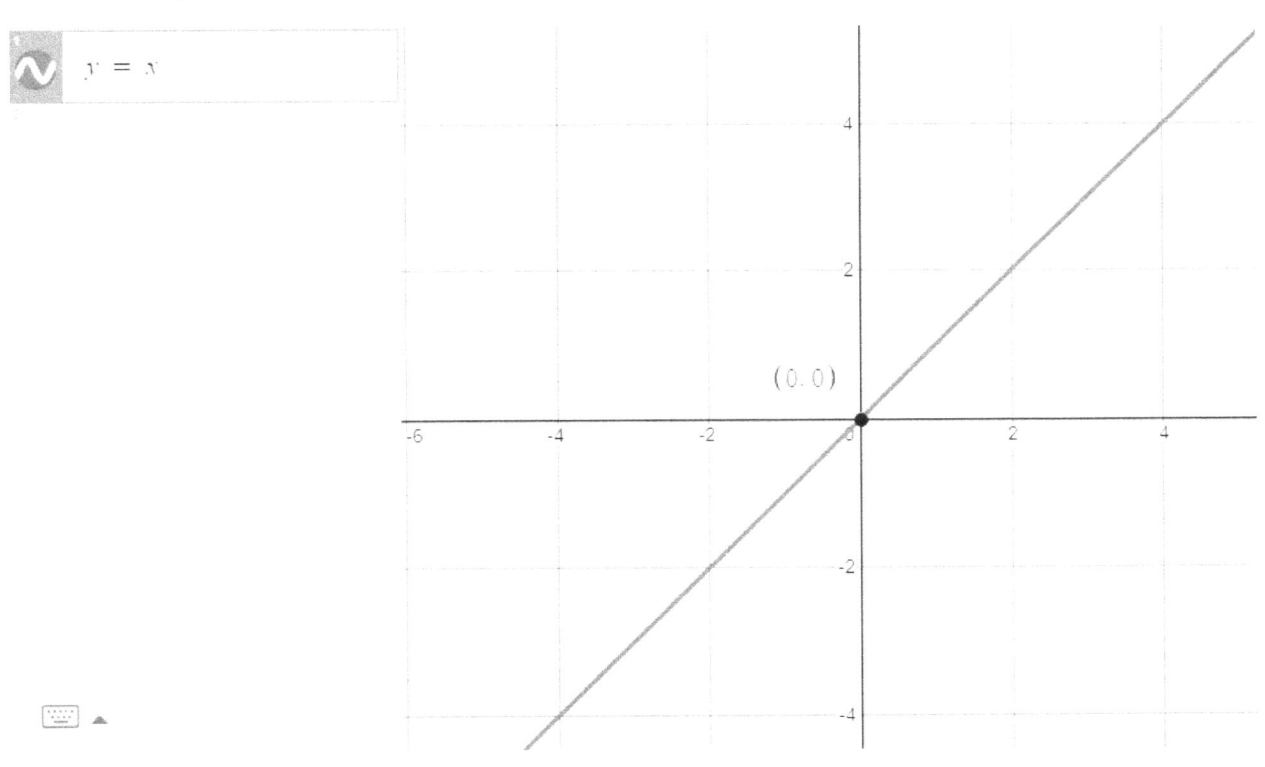

As you can see, in this case whatever you put in as x, will always be the resulting y-value. Another important feature to notice is that it passes directly through the origin (0,0).

We will now build a *Table of Values* for this equation:

x	y
-1	-1
0	0
1	1
2	2

The Slope-Intercept form of a Linear Equation

The most common form of an equation you will come across when dealing with Linear Equations is the **Slope-Intercept Form**. Here is how this form of equation is written, we will discuss each component separately after viewing:

$y = mx + b$

$m \text{ is the slope, } b \text{ is the } y - intercept$ where the line crosses

the $y - axis \text{ when } x = 0$), and x is the input value. You will now learn what slope is.

Slope is the rate of change, this can be determined by using the formula $\frac{rise}{run}$, *rise* is the y-coordinate and *run* is the x-coordinate. You pick a starting point, in this case (0,0) and you pick an end point such as (2,2). Rise in this case is 2 and the Run is also 2. $\frac{rise}{run} = \frac{2}{2} = 1, m = 1$, there is no y-intercept to add to the equation in this example (0 would not make a difference), the line passes through the origin, therefore; the equation is written like this $(y = mx) = (y = (1)x)$, $y = x$ is the final equation. Slope of line $y = x$ is 1.

Here is an example of a line that does **not** have a slope of 1:

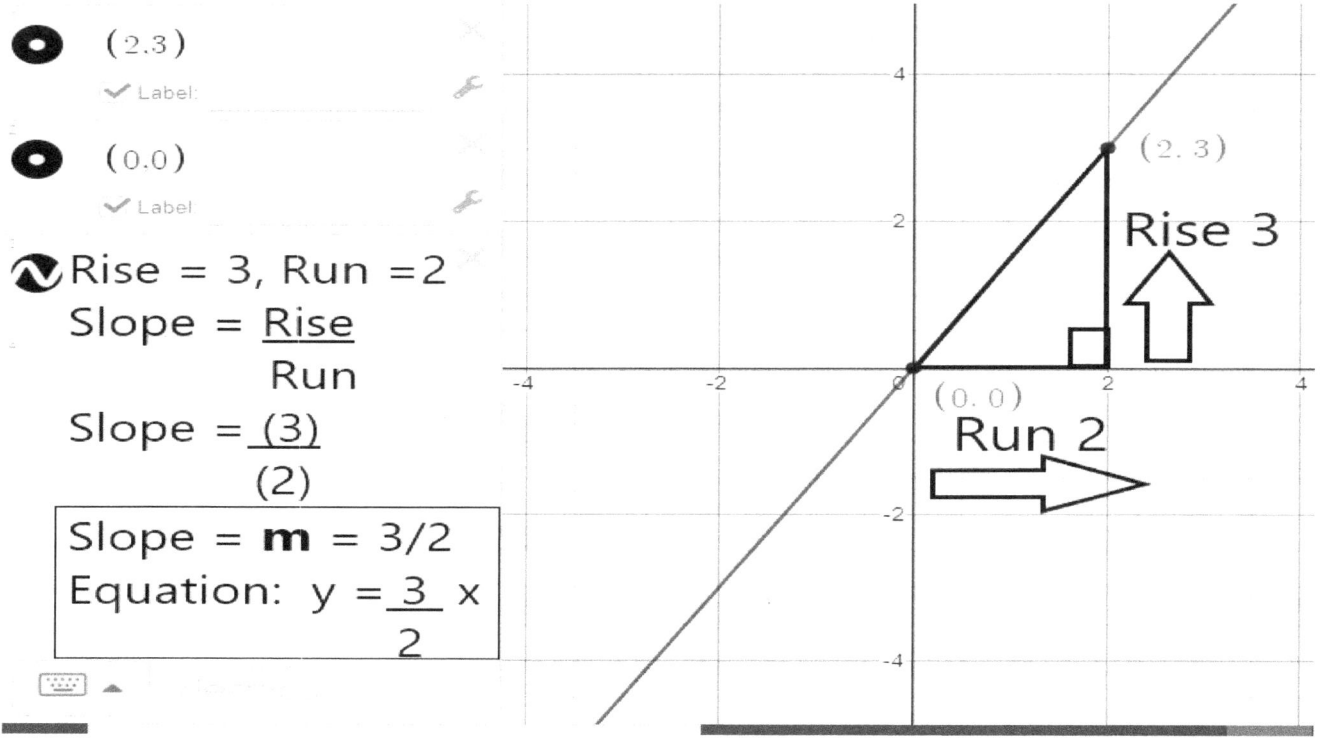

Slope of 3/2 means a rise of three units for every 2 units right.

What is the slope of this line?

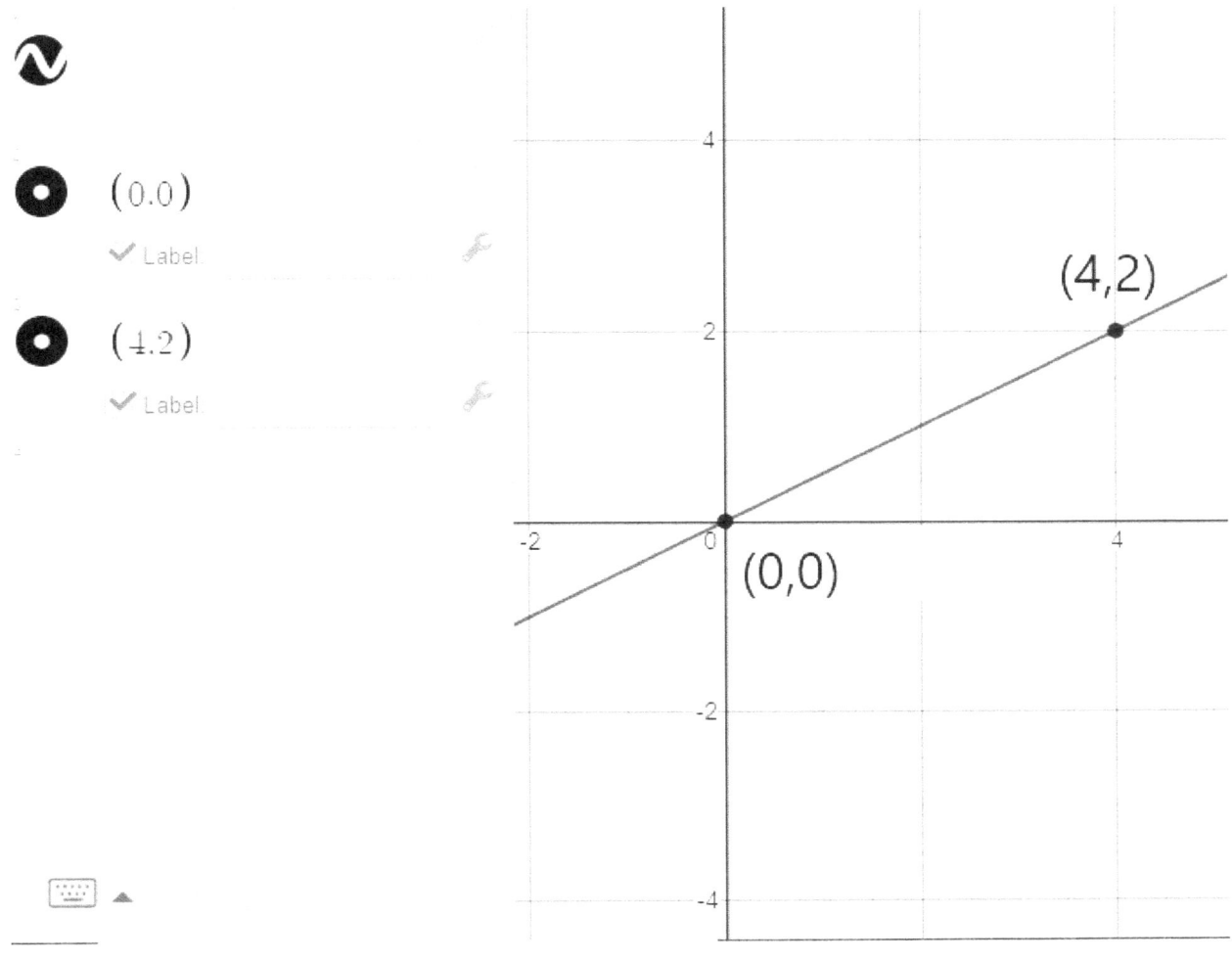

When you have figured out the slope please proceed to the next page:

You should have ½ as the slope. Rise was 2, run was 4, 2/4 = ½ m= ½

$$y = \frac{1}{2}x$$

Now you may be wondering, "What about when the line does not pass through the origin (0,0)?"

Well here is the formula:

$$\frac{y_2 - y_1}{x_2 - x_1}$$

This formula in my opinion is the most useful one, you pick two points a starting point and an end point, you subtract the y start point coordinate from the y end point coordinate and then you divide this by the x end point coordinate minus the x start point coordinate.

The graph below shows this formula being used:

- (2,2) (x_1, y_1)
  ✓ Label (2,2) Start point

- (4,3) (x_2, y_2)
  ✓ Label (4,3) End point

$\dfrac{y_2 - y_1}{x_2 - x_1} = \dfrac{3-2}{4-2}$

$= \dfrac{1}{2}$ Slope is 1/2
           m = 1/2

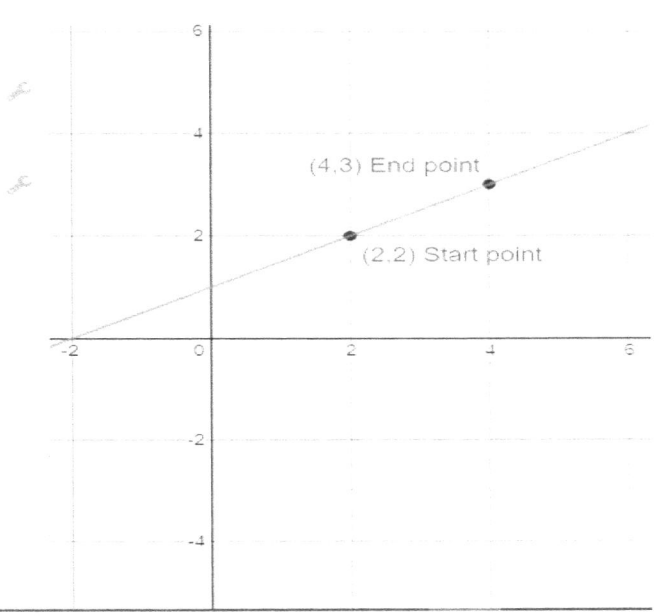

How do you find the y-intercept? Sometimes, you can look directly at the graph to see where the graph crosses the y-axis; however, that is not always the case, sometimes you must determine the y-intercept algebraically.

$\frac{1}{2}x + 1$

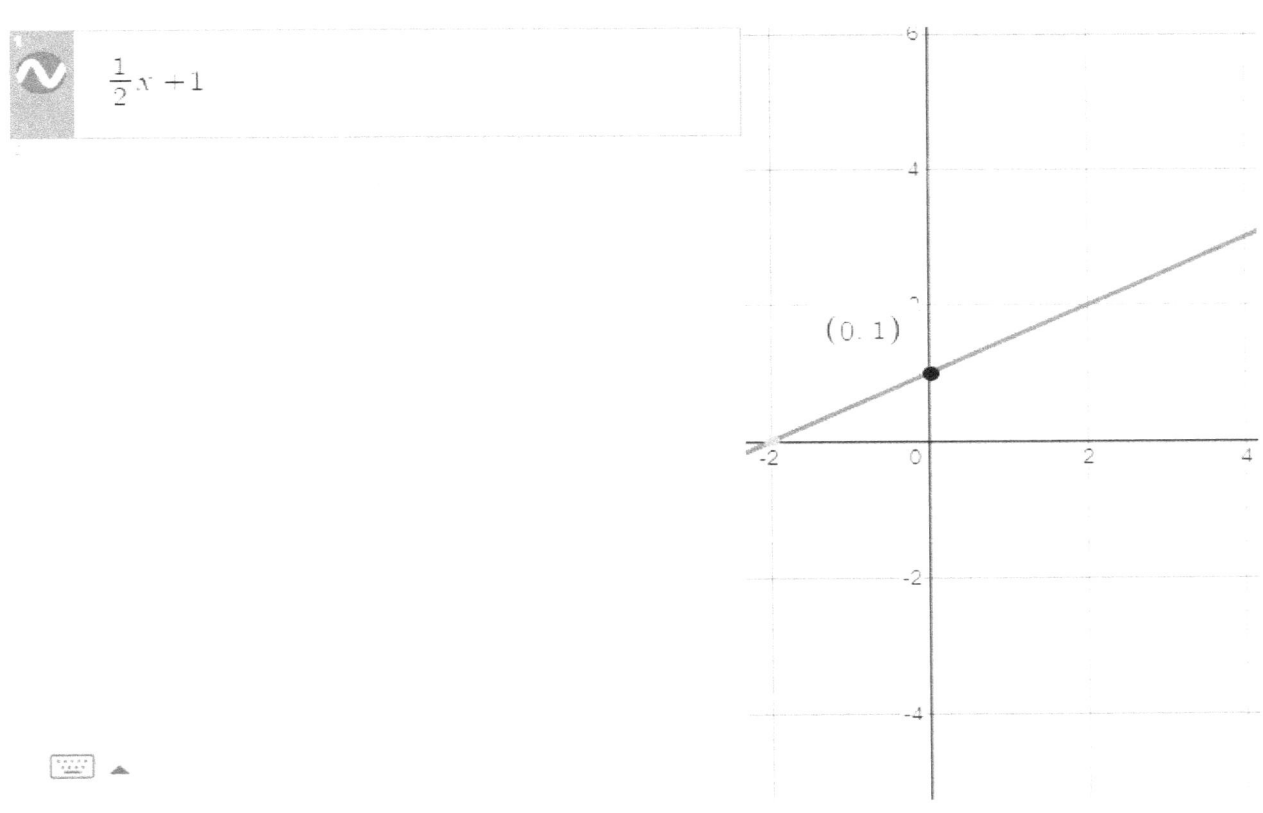

Writing a Linear Equation

In order to write a Linear Equation in the form $y = mx + b$, you need a few things. One, the slope (m). Two, the y-intercept (b-value). Sometimes you will be asked to write the equation according to its graph, other times you will have to write an equation according to the slope and a given point.

Example 1: A line with a slope of 5 has a y-intercept of 2, write the equation of this line.

Step 1: The slope is 5, therefore, m=5

Step 2: The y-intercept is 2, point (0,2) therefore, b=2

Step 3: Put this information together and substitute the two known values in the equation

$y = mx + b$ ⟶ m=5, b=2

$\boxed{y = 5x + 2}$ ⟵ $y = (5)x + (2)$

Example 2: A line with a with a y-intercept of 2, passes through the point (3,4), what is the equation of this line?

Since we are given two points, we can use the slope formula to determine the slope of this line.

Step 1: y-intercept is 2, point is (0,2); therefore, b=2, $(x_1 = 0, y_1 = 2)$

Step 2: Line passes through point (3,4), $(x_2 = 3, y_2 = 4)$

Step 3: "Plug" these numbers into the formula
$$m = \frac{y_2 - y_1}{x_2 - x_1}:$$

$$m = \frac{4-2}{3-0} = \frac{2}{3}$$

Slope is 2/3. Now that we have determined the slope and we know that the y-intercept is 2. We can finish writing the equation of the line.

$$m = \frac{2}{3}, b = 2$$

$$y = \left(\frac{2}{3}\right)x + (2)$$

$$y = \frac{2}{3}x + 2$$

X and Y-Intercepts

As you already know, the y-intercept is where the line crosses the y-axis. As you have probably noticed, there is also a part where the line crosses the x-axis, this is known as the x-intercept. To find the y-intercept you simply set x to 0 (x=0), to find the x-intercept, you make y=0.

Example 1: What is the y-intercept in the equation $y = 3x + 2$ *?*

Set x to 0: $y = 3(0) + 2$ $y = 0 + 2 = 2$

Y-intercept occurs at $x = 0, y = 2$, (0,2).

Example 2: What is the x-intercept in the equation $y = x + 3$ *?*

Set y to 0 (y=0)

$0 = x + 3$ so, now we see that x+3 =0, what does this mean? This means that we should isolate the x variable to solve for "x". We do this by rearranging the equation, exactly like you learned to do in *Chapter 1*.

$0 - 3 = x + 3 - 3$

$-3 = x$ when y=0, x=-3, point(-3,0), this is the x-intercept (where the line crosses the x-axis). Here is a visual:

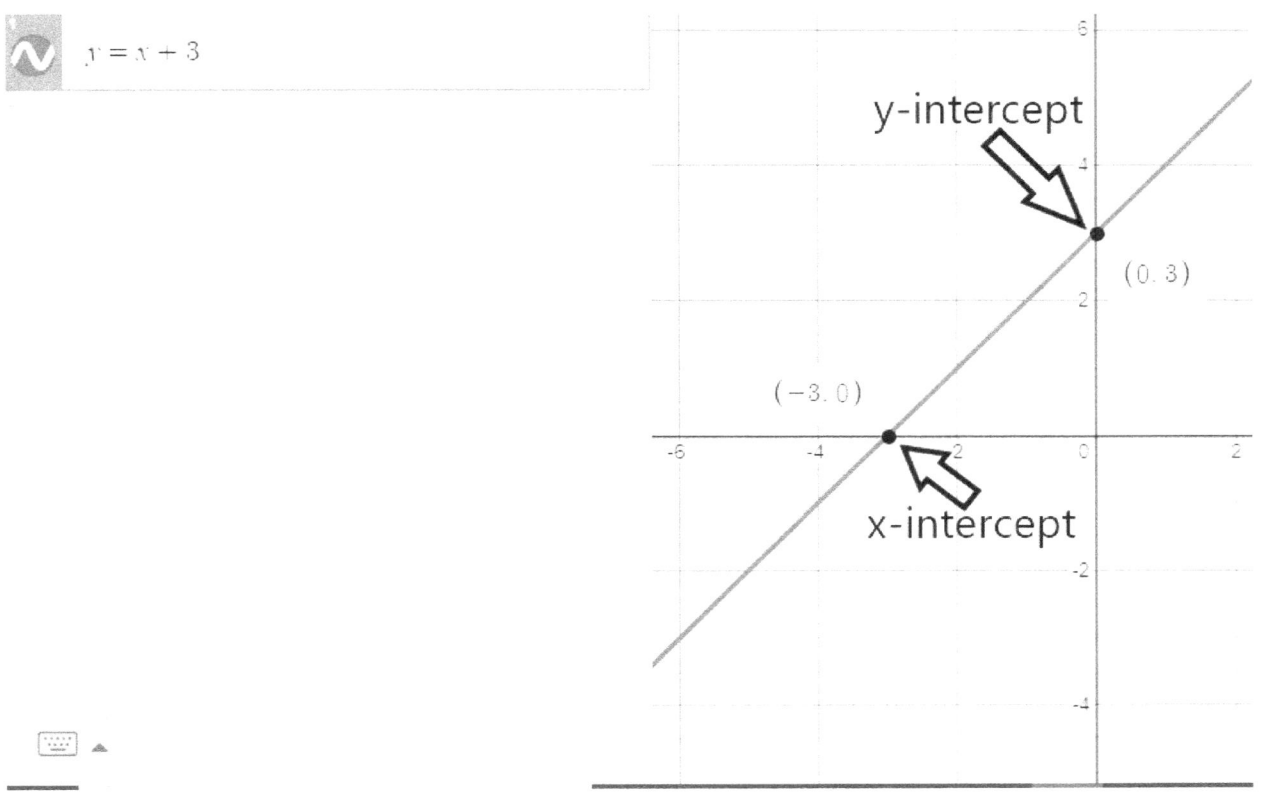

Modelling Equations Given a Situation

Sometimes you will be given a situation and you will be asked to determine an equation for it.

Example 1: **Jeremy wants to order pizza, the cost(C), in $, is determined by the number of Pizzas(n) and the one-time Delivery Charge of $10. It costs $20 per pizza. How much would it cost to order 3 pizzas?**

Step 1: Cost(C), is the dependent variable, so it basically represents "y" y=C

Step 2: The number of Pizzas(n) is basically the independent variable, so this represents (x), x=n

Step 3: It costs $20 per pizza; this is the slope(m).

Step 4: Delivery fee(D) of $10 is a one-time charge, this is the y-intercept. Whenever you see a statement like a base fee or a one-time charge, this will always be the y-intercept.

Step 5: Put this all together.

M = $20/Pizza

n=number of pizzas = 3

D = $10

$$C = mn + D$$

$$\boxed{C = (20)n + (10)} \longrightarrow \boxed{C = 20n + 10}$$

$n = 3,$

$C = \$20(3\ pizzas) + \$10 = \$60 + \$10 = \$70$

So if Jeremy wants to order 3 pizzas, it will cost him $70.

Functions

I will now introduce you to a special notation for graphs, usually you will come across this notation in your 2nd year in High School. This notation is known as *Function Notation*. What is a function? The formal definition of a function is, "A special kind of relation, in which for every Input there is only ONE output value." Here is a great example of what a Function is:

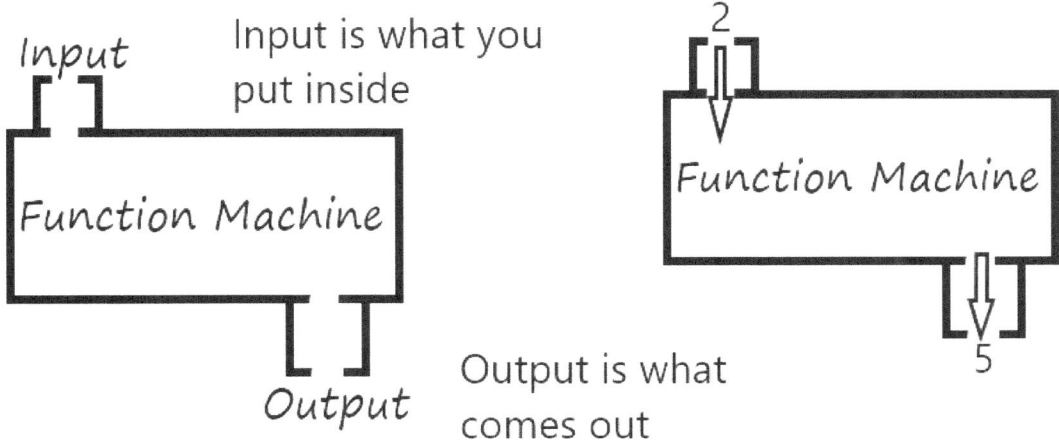

This is an important concept to grasp, what comes out of the function depends on what you put inside.

This is function notation:

$f(x) = x$

$f(x)\ is\ read\ as\ "f\ of\ x"$

The "f(x)" basically replaces "y" in the equation.

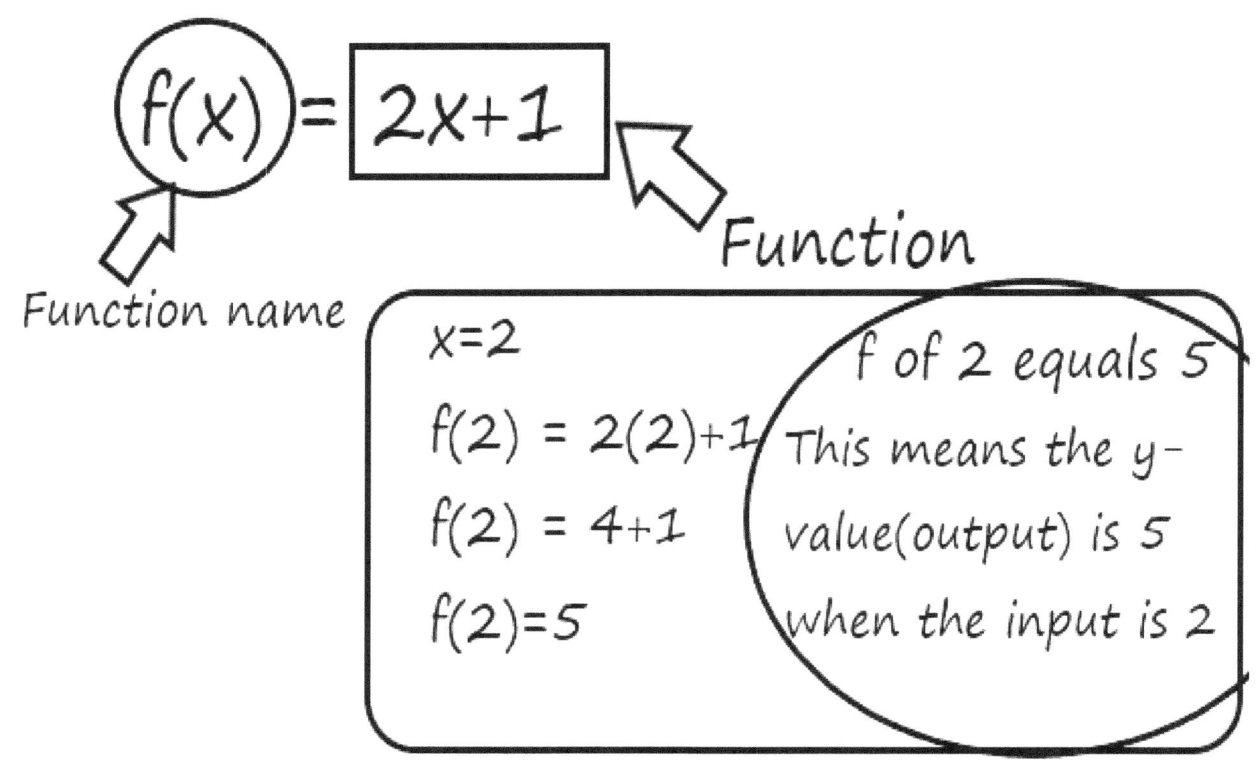

Corresponding point of this equation is (2,5)

When x=2, y=5

"f of 2 is 5"

Here is another example: $f(x) = -3x + 5$

Find $f(4)$. What this means is when x=4 what is the output.

Step 1: "Plug" in 4 as "x"

$f(4) = -3(4) + 5$

$f(4) = -12 + 5$

$f(4) = -7$ ⟶ $(4, -7)$

"F of 4 equals -7"

What would $f(x) = 8$ mean? This means that a certain input value results in what is on the other side of the equals sign, 8 (the output). How do we solve for x? We must plug in $f(x) = 8$.

$f(x) = -3x + 5$

$8 = -3x + 5$ solve for x,

$8 - 5 = -3x$

$3 = -3x$

$\dfrac{3}{-3} = -\dfrac{3x}{-3}$

$-1 = x$

$f(-1) = 8$ an input value of -1 results in an output value of 8.

Examples:

1. $f(x) = 5x^2 - 4$ $x = 3$
   Plug in $x = 3$, → $f(3) = 5(3)^2 - 4$
   $f(3) = 5(9) - 4$ → $f(3) = 45 - 4$
   $f(3) = 41$
   Coordinate $(3, 41)$

2. $f(x) = 2x + 3$ where $f(x) = -3$, find x

Set the left side equal to -3 (in other words replace $f(x)$ with -3)

$-3 = 2x + 3$ → $-3 - 3 = 2x$

$-6 = 2x$ → $-\dfrac{6}{2} = \dfrac{2x}{2}$ → -3

$x = -3$

Domain and Range

What is Domain and Range, well Domain is simply "all of defined x-values of a graph", Range is "all of the defined y-values of a graph." Sometimes you will have a set of ordered pairs and you will be asked to determine the domain and range, here is an example:

{(0,0),(1,1),(4,2),(9,3),(16,4)} *determine the domain and range.*

Well for the domain, we look at the first number in each coordinate set.

Domain: {0,1,4,9,16}

As for the range, we look at the second number in each coordinate set.

Range: {0,1,2,3,4}

Sometimes you will be asked to determine the domain and range of a function. Linear functions generally all have the same domain and range unless other wise stated or shown.

Domain: $\{x \mid X E R\}$ what does this mean? This means that "x is an element of all real numbers." To clarify, this means that x can have any input value.

Range: $\{y|YER\}$ what does this mean? This means that "y is an element of all real numbers." To clarify, this means that y can have any output value. Sometimes you will have another type of function with restrictions on certain input values, we will not be learning about this in this book as this is not covered until Grade 11.

Example 1: What is the domain and range of $f(x) = x + 3$?

The domain is, $\{x|xER\}$ all real numbers

The range is, $\{y|yER\}$ all real numbers

This will always be the case, unless you are given set of ordered pairs.

To repeat, the domain and range will always be "All Real Numbers" unless you are given a certain set of ordered pairs.

Practice Questions

1. Which Quadrants do the following statements describe?
   a) x and y are both negative.
   b) x and y are both positive.
   c) x is negative, y is positive.
   d) x is positive, y is negative
2. Plot the point (1,4)

3. Write the equation of a line with a slope of 3, and a y-intercept of 7. Hint: Write it in the form $y = mx + b$

4. What is the slope of this line? Point 1: (3,4),   Point 2: (7,6).

5. What is the y-intercept of $y - 7 = x$?
6. What is the x-intercept of $y = x - 3$?

7. If $f(x) = 2x + 2$, What is $f(2)$?

8. What is the Domain and Range of $f(x) = x - 7$?

9. What is the Domain and Range of {(1,2),(3,4),(5,6),(7,8)}?

TRIGONOMETRY (Introduction)
Chapter 4

Trigonometry is a branch of mathematics that studies relationships involving lengths and angles of triangles. It uses ratios and applications of the *Pythagorean Theorem* to solve problems.

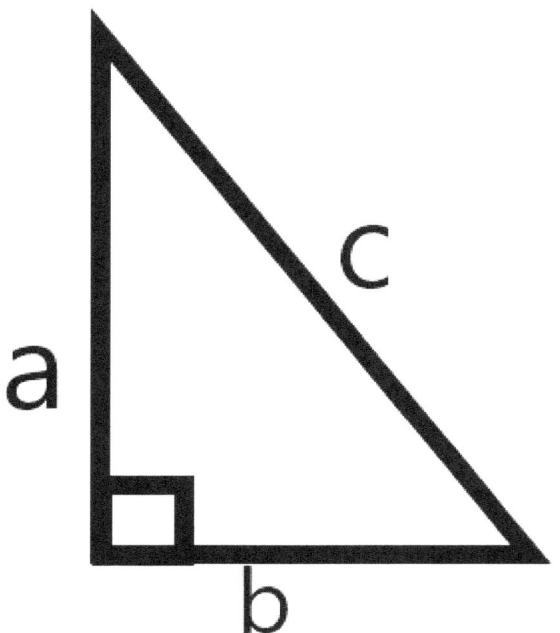

Right Triangles:

- Have one angle that equals 90°
- Are like other triangles:
    - Have 3 angles
    - If you add up all the Angles it equals 180°

Pythagorean Theorem:

- It applies only to right triangles.
- The sum of the squares of the legs equals the hypotenuse squared:
$$c^2 = a^2 + b^2$$

Example 1: Find the length of side *c*.

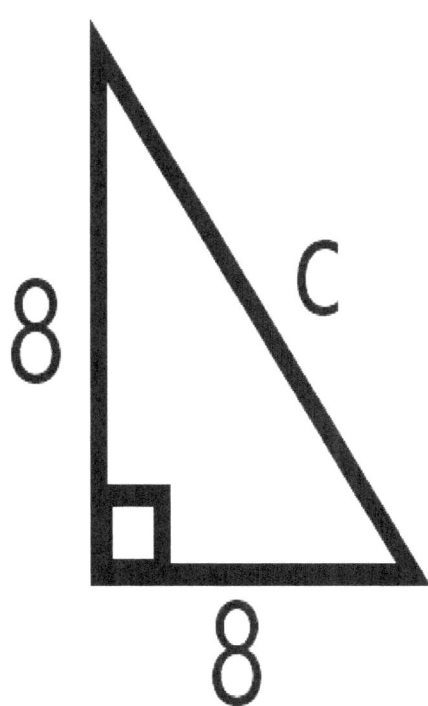

$$a^2 + b^2 = c^2$$
$$8^2 + 8^2 = c^2$$
$$64 + 64 = c^2$$
$$128 = c^2$$
$$\sqrt{128} = c$$
$$11.3 = c$$

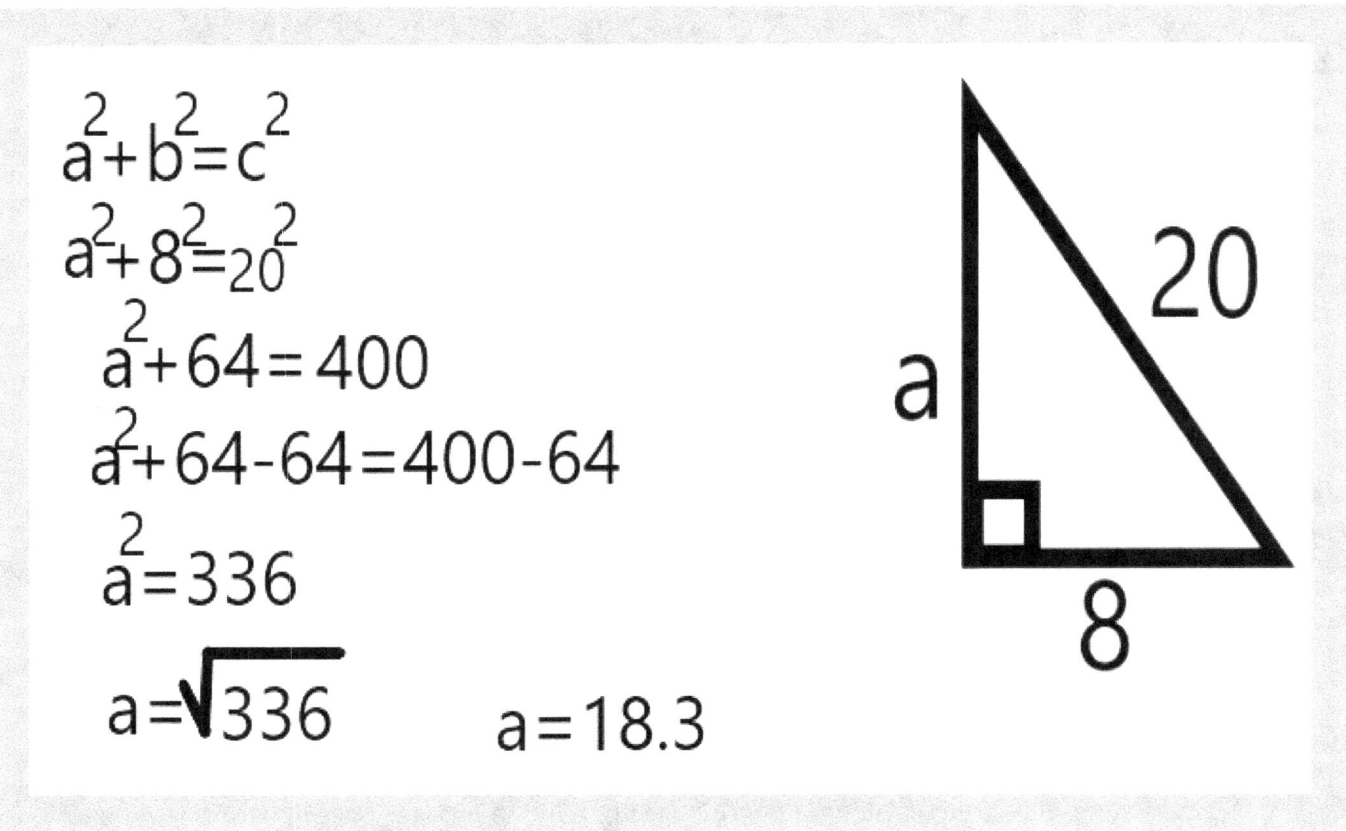

$a^2 + b^2 = c^2$

$a^2 + 8^2 = 20^2$

$a^2 + 64 = 400$

$a^2 + 64 - 64 = 400 - 64$

$a^2 = 336$

$a = \sqrt{336}$ $a = 18.3$

Example 2: Find the length of side *a,*

Labelling a Triangle:

- The angle is always opposite of the side (across from it).
- Angles are spelt with capital letters
- Sides are spelt in lower-case letters

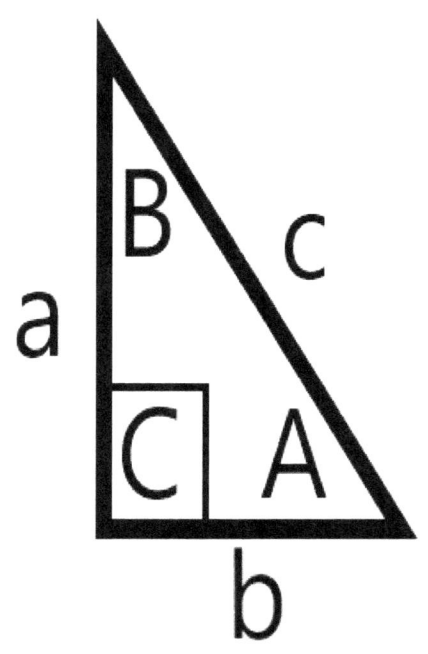

Trigonometry started with the ancient Greeks, people like Euclid, Pythagoras, Archimedes, and many others studied triangles and the relationship between their angles and sides. Recall that for a right angle triangle we label the sides with a, b, and c and we label angles with A, B, and C. Note that <A (angle A) is the angle opposite from side "a" and that <B (angle B) is opposite from side b.

Here is a way to represent all the sides in respect to <A:

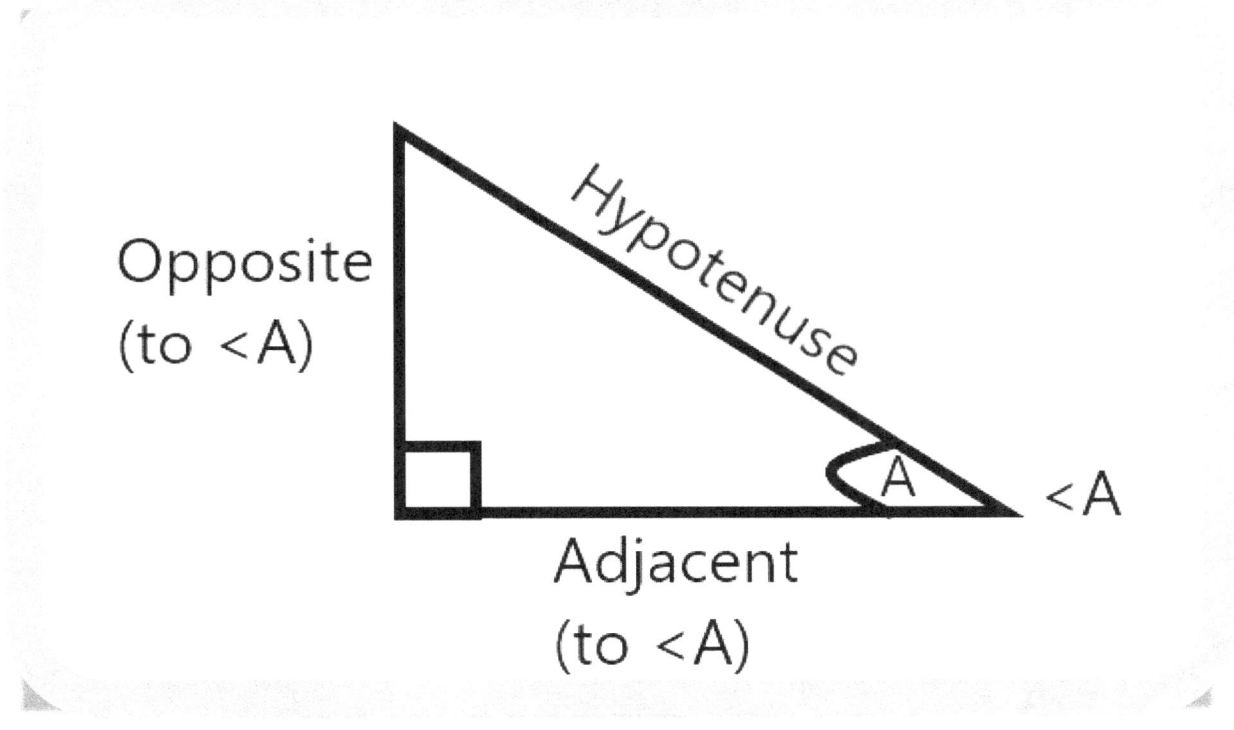

Adjacent means "beside" and opposite means "across from."

The ancient Greeks also noticed that for the angles of a right triangle the ratios of

$$\frac{opposite}{adjacent}, \frac{opposite}{hypotenuse}, \frac{adjacent}{hypotenuse}$$

These ratios were <u>always the same</u> for the <u>same angle</u>

To illustrate, consider the following two triangles that have the same angle A that in this case, equals 34°. The ratios are shown below,

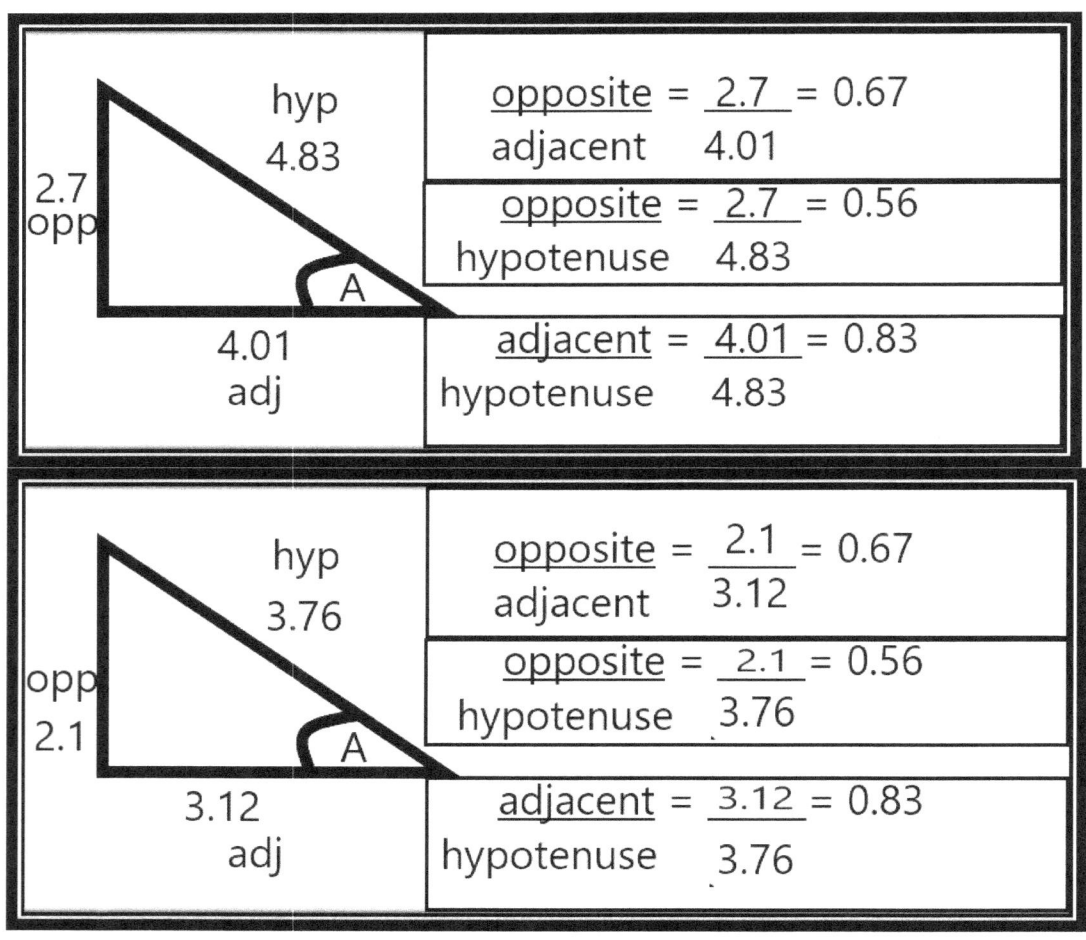

The important thing to grasp is that the same ratio value emerges for the same angle no matter how large or how small the triangle.

Here are the three primary trigonometric ratios:

$$sine = \frac{opposite}{hypotenuse}$$

$$cosine = \frac{adjacent}{hypotenuse}$$

$$tangent = \frac{opposite}{adjacent}$$

Here is a mnemonic to remember this, SOH CAH TOA.

Before moving on, ensure your calculator is in Degrees mode. Sometimes your calculator will be in Radians mode. For now, we will be focusing on degrees.

Tangent Ratio

$$tangent = \frac{opposite}{adjacent}$$

There are some examples of the Tangent ratio on the next page,

Tan 27° = 0.5095   Tan 45° = 1   Tan 57° = 1.5399

If given the ratio we can find the angle.

Example 1: Given, Tan A = 0.8391 find angle A.

In order to do this, we must use the inverse tan function ($\tan^{-1}$) you can do this by pressing the 2nd function key on your calculator and then you press "tan."

$$A = \tan^{-1}(0.8391)$$
$$A = 40°$$

Try to get these answers yourself:

- Tan A = 0.5774   A = 30°
- Tan B = 0.5095   B = 27°
- Tan C = 1   C = 45°
- Tan D = 1.5399   D = 57°

Example 2: Determine angle B.

Tan B = Opposite / Adjacent

Tan B = CA / BC

Tan B = 1.25

B = Tan⁻¹(1.25)

B = 51°

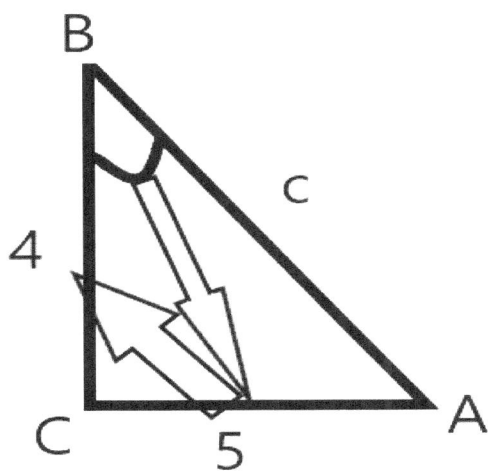

As you can see, the Tangent ratio is equal to the length of the opposite side divided by the length of the adjacent side.

The Sine and Cosine Ratio

$$sine = \frac{opposite}{hypotenuse}$$

$$cosine = \frac{adjacent}{hypotenuse}$$

The sine and cosine ratios are very similar to the tangent ratio. As we saw above, the sine of an angle is the ratio of the opposite over the hypotenuse, while the cosine of an angle is the ratio of the adjacent over the hypotenuse.

To illustrate the use of the sine ratio, consider <K in the triangle below. Let us calculate angle K to one decimal place. We look to see what sides are in relation to K. 4.3 is the hypotenuse and 4.0 is the opposite from K.

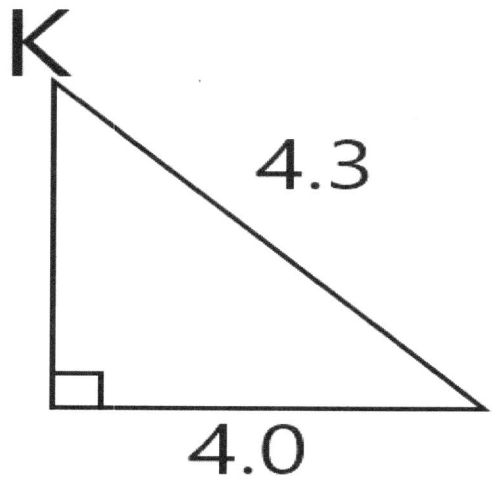

The function that involves opposite and hypotenuse is the sine function. Therefore:

$$\text{Sin } K = \frac{opp}{hyp} = \frac{4.0}{4.3} = 0.9302$$

$K = \sin^{-1}(0.9302)$

$K = 68.5°$

Below is an example of the cosine ratio being used,

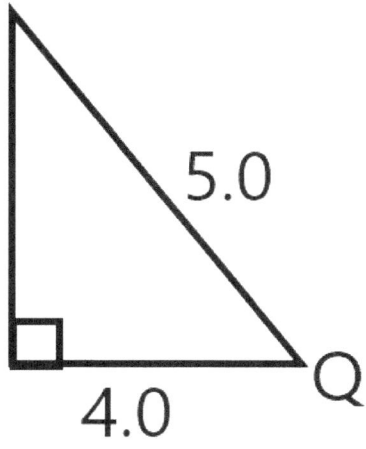

$$\text{Cos } Q = \frac{4}{5} = 0.8$$

$$\cos^{-1}(0.8) = 36.9°$$

***Remember that $\cos = \dfrac{adjacent}{hypotenuse}$ adjacent side was 4, and the hypotenuse was 5. We divide 4 by 5, and then we use this ratio to determine the angle. We find the angle by using the $\cos^{-1}()$ function on the calculator.

Now that we know how to find and angle using a trigonometric ratio, we will now look at how to determine the length of a side by using one of the primary trigonometric ratios. After that we will look at how to apply a trig ratio in a real-life situation.

Example 1: Determine a, to the nearest tenth of a centimeter. ***<C is 45.6 degrees***

Cos C = adj
         hyp

Cos 45.6° = adj
             7.2

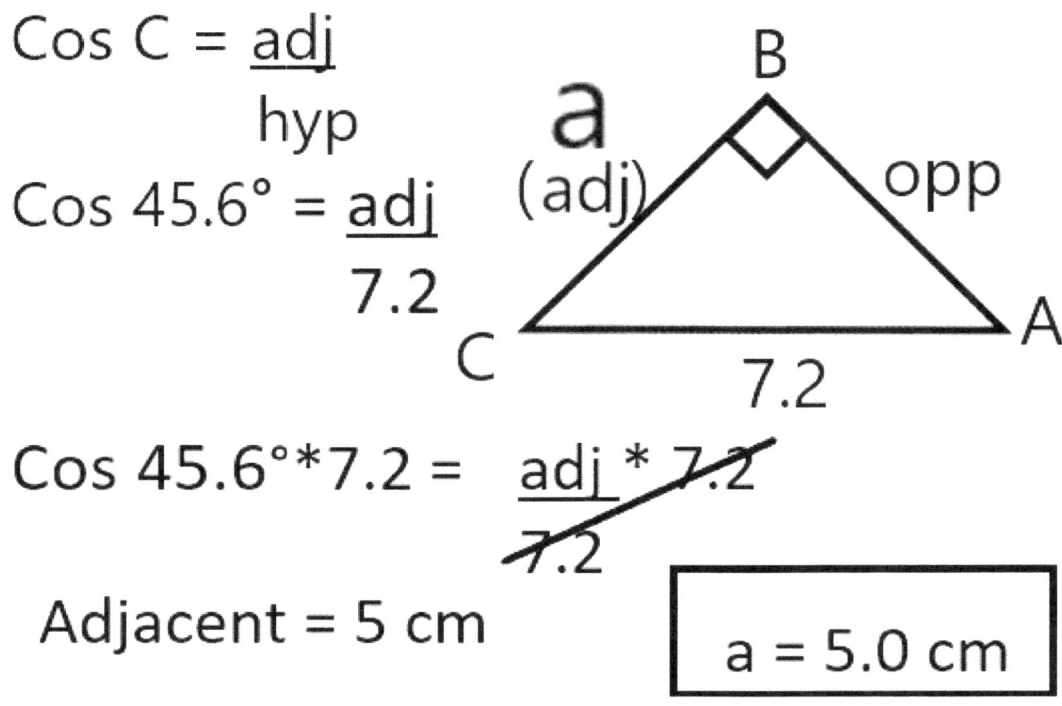

Cos 45.6°*7.2 = adj * 7.2 / 7.2

Adjacent = 5 cm

a = 5.0 cm

Example 2: If $\sin 25° = \dfrac{1}{x}$ what is the value of x?

Step 1: What do we need to find? *We need to find the length of the hypotenuse. Let's assume that the length of the opposite side is '1'*

Step 2: Solve for x.

$\sin 25° * x = \dfrac{1}{x} * x$ multiply both sides by x to eliminate it from the right side.

$$\sin 25° * x = 1$$

$$\frac{\sin 25° * x}{\sin 25°} = \frac{1}{\sin 25°}$$ Divide both sides by $\sin 25°$

$$x = \frac{1}{\sin 25°} \qquad x \approx 2.37$$

As I have mentioned earlier, we will now look at applying a trig ratio in a real-life situation.

Example 3: A guy wire supporting a cell tower is 24m long. If the wire is attached at a height of 17m up the tower, determine the angle that the guy wire forms with the ground.

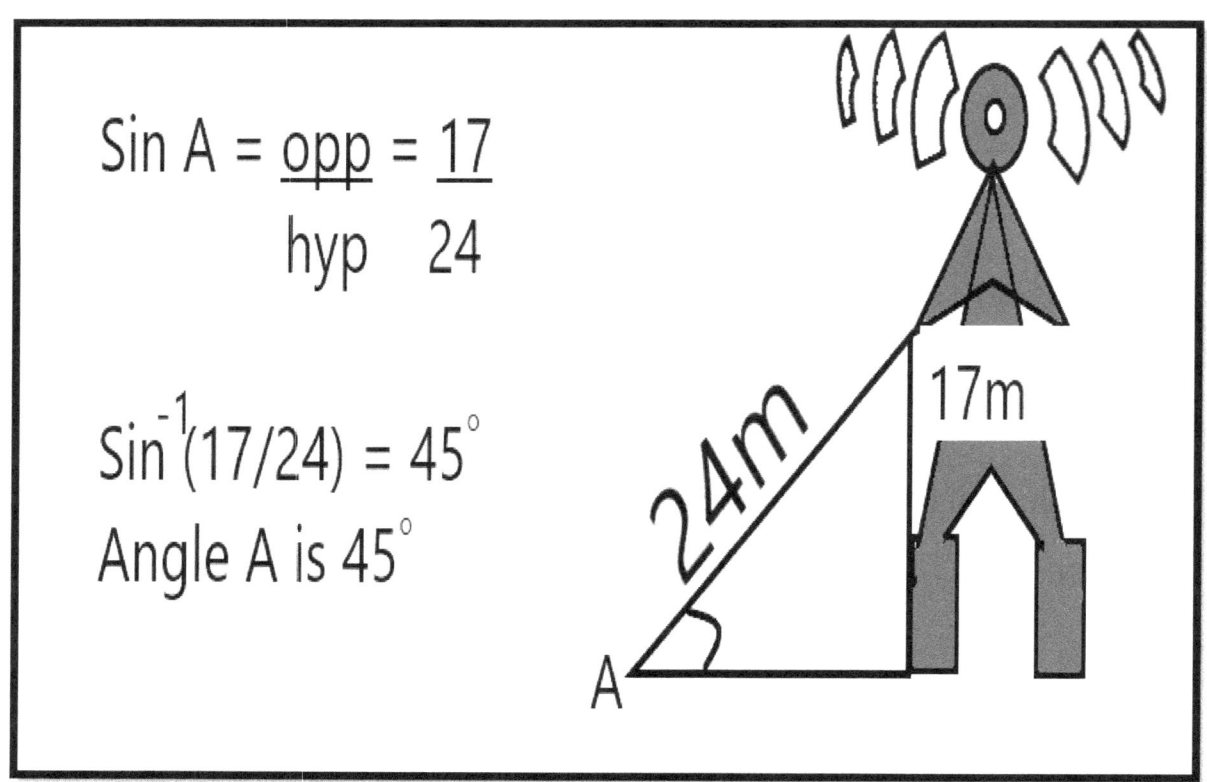

The height of the triangle we formed is 17 meters.

The length of the guy wire is 24 meters, this is the hypotenuse.

Using this information, we can use the Sine ratio ($\sin A = \dfrac{opp}{hyp} = \dfrac{17}{24}$)

Once we have the sine ratio ($\dfrac{17}{24}$), we use the inverse sine function $\sin^{-1}()$.

This gives us an angle of approximately $45°$.

Practice Questions

1. For the indicated angle in the triangle below give the sin, cos and tan ratios AND the value of the angle.

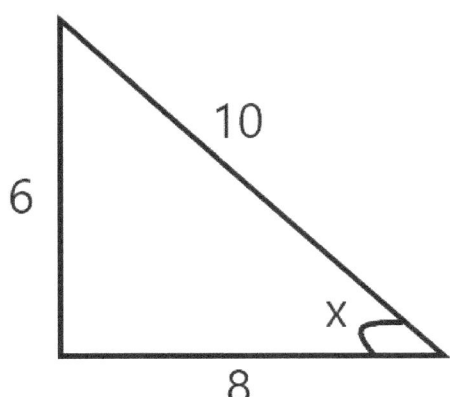

2. Find Angle A. Round your answer to the nearest tenth.

a)

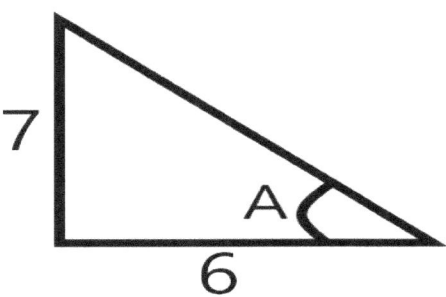

b)

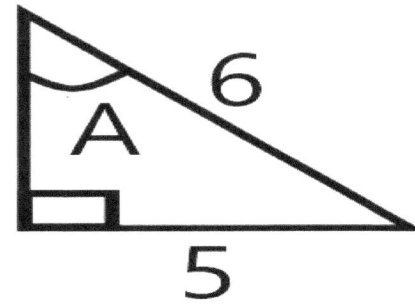

3. If $\sin 45° = \dfrac{6}{x}$, what is the value of x?

4. A ladder is leaning against the top of a wall. The wall is 4 meters high. The ladder is 7 meters long. How far is the base of the ladder from the wall (length b)? What is the measure of the angle that is formed between the bottom of the ladder and the ground (angle pointing towards the wall)?

5. Find the value of y. Round your answer to 2 decimal places.

   a) $\tan 26° = \dfrac{y}{2}$

   b) $\cos(y) = \dfrac{1}{2}$

   c) $\sin 35° = \dfrac{6}{y}$

   d)

6. Determine the missing angle and the missing side lengths for the triangle below.

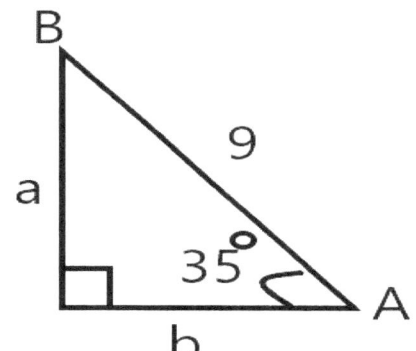

Angle B = ?

a = ?

b = ?

FACTORING POLYNOMIALS

Chapter 5

We will begin this chapter by learning the different classifications of numbers.

In mathematics, numbers are classified according to common characteristics. Every number is classified as belonging to <u>one or more</u> of the following sets of numbers:

- Natural Numbers: N= {1,2,3, ...}
- Whole Numbers: W= {0,1,2,3, ...}
- Integers: $I = \{..., -3, -2, -1, 0, 1, 2, 3, ...\}$
- Rational Numbers: $Q = \{\frac{a}{b} | a, b \in I, b \neq 0\}$
- Irrational Numbers:
$\bar{Q} = \{non-terminating, non-repeating\ decimals\}$
- Real Numbers: R= {all rational & irrational}

Factors and Products

Factor:

- Is a number that is written as a product
- *Example: 6 = 3*2*

3 and 2 are the factors

Here are some more examples:

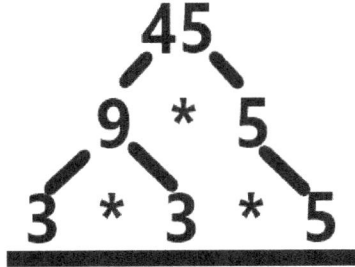
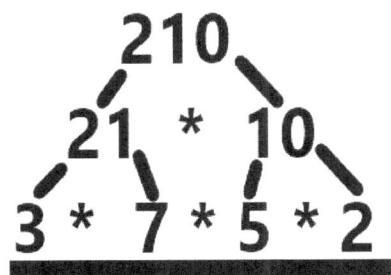

Prime Number:

- Number greater than 1
- Its only factor is itself and 1
- Example: 2, 3, 5, 7, 11, 13, 17, 19, 23, 29, 31...
- The number 1 is neither prime nor composite

Greatest Common Factor (GCF)

GCF is the greatest integer that is a factor of each number. It can also be considered the largest number that can divide into all the numbers.

Example 1: What is the GCF of *35, 15, and 25*?

*35 = **5***7*

*15 = 3***5***

*25 = **5***5*

As you can see the GCF of these 3 numbers is 5.

Example 2: Find the GCF of 42 and 60

- Step 1: Write the prime factors of each numbers:
$$42 = 2 * 3 * 7$$
$$60 = 2 * 2 * 3 * 5$$
- Step 2: Find the prime factors. The common prime factors as you can see are 2 and 3.
- Step 3: Multiply the common prime factors.
GCF = 2*3 = 6
6 is the largest factor of both numbers

GCF of Monomials

To find the GCF of monomials, you do the same thing you did in the section before, but with a few twists:

$$a^2 = a * a$$

$$a^2 b^3 = a * a * b * b * b$$

GCF of these two monomials is a^2

Example: find the GCF of $40a^2 b$ & $48ab^4$.

- Step 1: Write the prime factorization,
$$40a^2b = 2 * 2 * 2 * 5 * a * a * b$$
$$48ab^4 = 2 * 2 * 2 * 2 * 3 * a * b * b * b * b$$

- Step 2: Multiply those numbers numbers/variables together.
  The GCF is 2*2*2*a*b = 8ab

Example: find the GCF of $4x^2y^3$ & $6xy^2$.

- Step 1: Write the prime factorization,
$$4x^2y^3 = 2 * 2 * x * x * y * y * y$$
$$6xy^2 = 2 * 3 * x * y * y$$

- Step 2: Multiply those numbers numbers/variables together.
  The GCF is $2 * x * y * y = 2xy^2$

Factoring Binomials

Now that we have looked at determining the GCF of monomials, we will now look at how to factor binomial expressions (a two-term expression).

Example: Factor $25a^2 + 15a$

- Step 1: Find the GCF for the two terms: factor each term first, and multiply everything that is not in the circles together for each term,
$$25a^2 = \boxed{5} * 5 * \boxed{a} * a = 5a$$
$$15a = \boxed{5} * 3 * \boxed{a} = 3$$

- Step 2: GCF = $5 * a = 5a$
- Step 3: Place the remainder of each term inside the brackets and put the GCF outside of the brackets like this:
$5a(5a + 3)$
- Step 4: Verify by distributing,
$$5a(5a + 3) = 25a^2 + 15a$$

Example: Factor $18x^2 - 12x^3$

- Step 1: Find the GCF for the two terms: factor each term first, and multiply everything that is not in the circles together for each term,

$$18x^2 = 2 * 3 * 3 * x * x = 3$$
$$12x^3 = 2 * 2 * 3 * x * x * x = 2x$$

- Step 2: GCF = $2 * 3 * x * x = 6x^2$
- Step 3: Place the remainder of each term inside the brackets and put the GCF outside of the brackets like this:
$6x^2(3 - 2x)$
- Step 4: Verify by distributing,

$6x^2(3 - 2x) = 18x^2 - 12x^3$

Now its your turn, try these two practice questions, the answer to these two questions will be on the next page.

a) $6n + 9$

b) $6c + 4c^2$

a) $6n + 9 = 3(2n + 3)$
b) $6c + 4c^2 = 2c(3 + 2c)$

Multiplying Binomials

(The Foil method)

The distributive property:

For any real numbers a, b, & c:

$$a(b + c) = ab + ac$$

Example: $8(x + 3) = 8x + 24$

When you multiply two binomials, the only difference is that there will be two terms out front instead of one.

To simplify this type of problem, you just need to distribute twice.

Example: multiply $(2x + 3)(5x + 8)$

$$= 2x(5x + 8) + 3(5x + 8)$$

$$= 10x^2 + 16x + 15x + 24$$

Combine like terms: (16x and 15x)

$= 10x^2 + 31x + 24$

Factoring Polynomials of the form $ax^2 + bx + c$ when $a = 1$

Example: Factor $x^2 + 6x + 8$

    First, check for common factor(s). Do we have a GCF? No!

So what do we do? We look for a number that multiplies to the constant term (8) and the sum of these two numbers is the middle coefficient (6)

(*) 8

(+) 6

2 & 4

2 and 4 multiplies to 8 and adds up to 6. Once we have figured out these two numbers, we put the first number in the brackets with the x like this, $(x + 2)$

And then we put the second number in another pair of brackets with the x $(x + 4)$, and then we multiply the two pair of brackets with each other like this:

(x+2)(x+4)

$$x^2 + 6x + 8 = (x+2)(x+4)$$

The method above for factoring polynomials of this form, will only work if the coefficient in front of the x^2 is 1.

Example: Factor $x^2 - x - 12$

   First, check for common factor(s). Do we have a GCF? No!

So what do we do? We look for a number that multiplies to the constant term (-12) and the sum of these two numbers is the middle coefficient (-1)

(*) -12

(+) -1

3 & -4

3 and -4 multiplies to -12 and adds up to -1. Once we have figured out these two numbers, we put the first number in the brackets with the x like this, $(x+3)$

And then we put the second number in another pair of brackets with the x like this, $(x-4)$, and then we multiply the two pair of brackets with each other like this:

(x+3)(x-4)

$$x^2 - x - 12 = (x+3)(x-4)$$

The method above for factoring polynomials of this form, will only work if the coefficient in front of the x^2 is 1.

In the next section we will look at when the coefficient in front of the x^2 is greater than 1 (a>1)

Factoring Polynomials of the form $ax^2 + bx + c$ when $a > 1$

When you factor Polynomials of the form $ax^2 + bx + c$ when $a > 1$, there are a few extra steps. The best way to show you is by providing an example,

Example: Factor $2x^2 + 11x + 12$,

- Step 1: Since the number in front of the x^2 term is greater than 1, then we basically find two numbers that multiply to equal the constant term (term without variable, so in this case "12") times the "a" value (a=2). Therefore, 12*2 =24, you need to find two numbers that multiply to 24 and add up to 11.

(*) 24
(+) 11
3 & 8

3 times 8 is 24, and 3 plus 8 is 11. This is very important, these two numbers become the decomposition of the middle term.

- Step 2: 11x becomes 3x +8x.
$$2x^2 + 8x + 3x + 12$$
- Step 3: Now that we have four terms, we now group the terms in pairs of two.
$$[2x^2 + 8x] + [3x + 12]$$
- Step 4: Find the GCF of each pair. The terms you get in the brackets of both pairs should be the same. You will see why in just a moment…

GCF of $2x^2$ and $8x$ is $2x$

$$2x^2 = 2 * x * x = x$$
$$8x = 2 * 2 * 2 * x = + 4$$

Remember "x" and "+4" is the remainder after taking 2x out of both terms.

GCF of $3x$ and 12 is 3

$3x = 3 * x = x$
$12 = 3 * 2 * 2 = + 4$

Did you notice how the remainder for both pairs was exactly the same? This is what we are looking for when you factor by grouping

- Step 5: Remove the GCF of both pairs and place the remainder of both pairs in the brackets.
  $2x(x + 4) + 3(x + 4)$
  GCF of first pair goes outside of the first brackets, GCF of second pair goes outside of the second pair of brackets. So, what do we do now? You will see the what to do next in "Step 6"
- Step 6:
  $GCF\ of\ both\ pairs\ goes\ together\ in\ the\ brackets\ like\ this$:
  $$(2x + 3)$$
  Then you keep (x+4) in the brackets and put it with (2x+3); therefore,
  $2x^2 + 11x + 12 = (2x + 3)(x + 4)$
- Step 7: Verify your answer,
  $2x(x + 4) + 3(x + 4) = 2x^2 + 8x + 3x + 12$
  $= 2x^2 + 11x + 12$

The final answer is correct. With this being said, $(2x + 3)(x + 4)$ are the factors of $2x^2 + 11x + 12$. You will learn in Grade 11, why factoring is important. Basically, factoring is used to determine the "zeros" of an equation (where the graph crosses the x-axis, this is also known as the x-intercepts). Factoring will also be used to simplify rational expressions when there is a polynomial in the numerator and in the denominator.

You have now finished reading Chapter 5, you may now try the practice questions.

Practice Questions

1. Write each number as a product of its prime factors.
    a) 60
    b) 88
    c) 390

2. What is the GCF of the following number pair?
   48 and 72?

3. Factor each polynomial using common factoring.
   a) $2x^2 + 4x$
   b) $3x + 3$
   c) $5x^2 - 15x$
   d) $y^2 + y$
   e) $2ab + b$

4. Factor these polynomials using the method shown in "Factoring Polynomials of the form $ax^2 + bx + c$ when a=1"
   a) $x^2 + 11x + 30$
   b) $y^2 + 17y + 60$
   c) $x^2 - x - 6$

5. Factor these polynomials using the method shown in "Factoring Polynomials of the form $ax^2 + bx + c$ when a>1"
   a) $5x^2 - 7x + 2$
   b) $6x^2 + 11x - 10$

VOLUME & SURFACE AREA OF GEOMETRIC FIGURES

Chapter 6

In this chapter, we will be learning about the Volume and Surface Area of Geometric Figures.

Volume

What is Volume? Volume simply put; is the amount of 3-Dimensional space an object takes up.

We will begin by looking at the most basic 3-dimensional shape, the Cube.

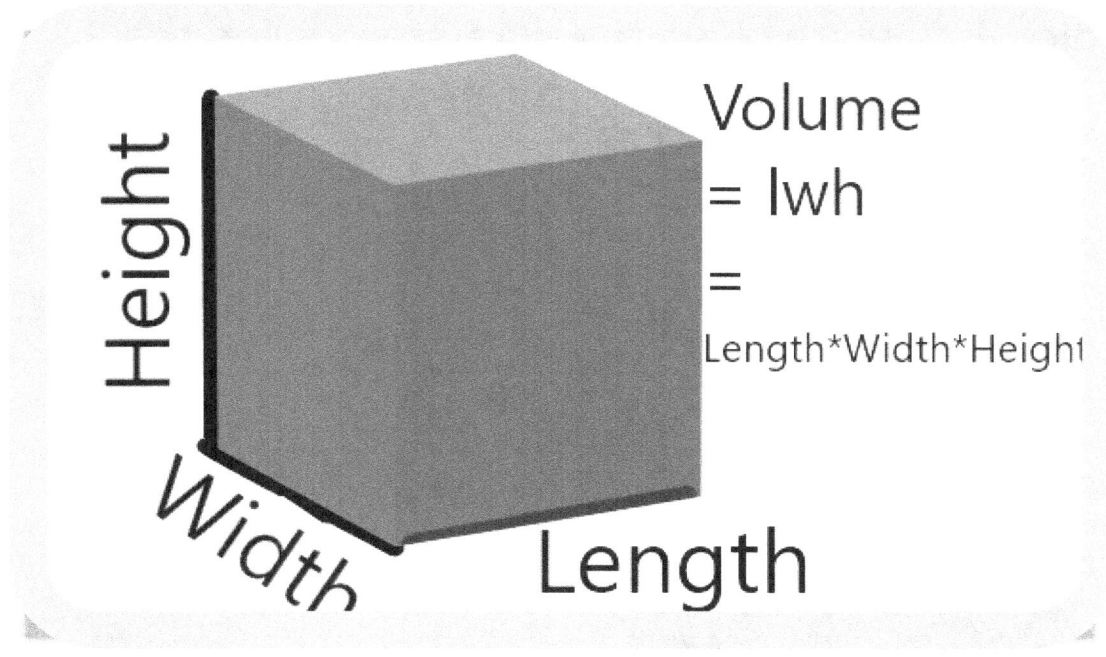

All dimensions of a Perfect Cube are always the same. For example, if the length is 7cm, then the width and height is also 7cm. If the width is 60cm, then the length and height is 60cm.

To determine the volume of a sphere, we use the formula:

$$V = \frac{4\pi r^3}{3}$$

r is the Radius. Radius is the distance from the centre of the circle to any outside point of the circle. In other words, this is "Half-way across." Sometimes you will be given the diameter of a sphere. Diameter is the distance between any two points on a circle that passes through the centre point. If you are given the diameter, divide the diameter by 2 to get the radius. $r = \dfrac{diameter}{2}$

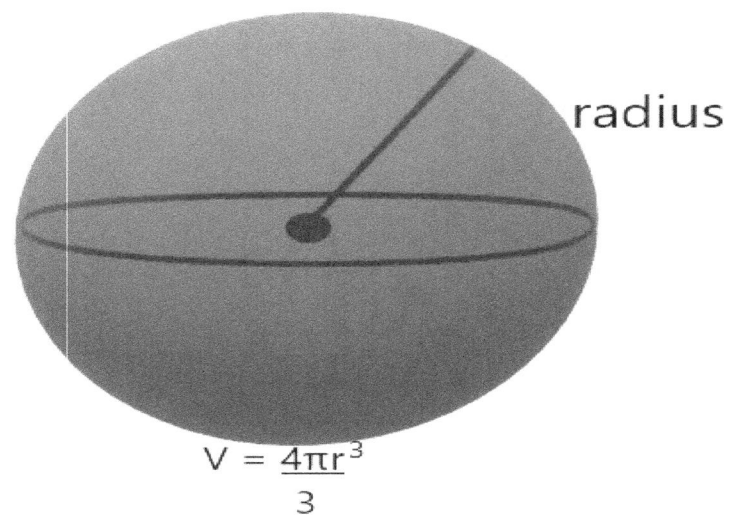

$V = \dfrac{4\pi r^3}{3}$

Find the Volume of a Sphere that has a radius of 2 inches:

The radius is 2 inches, therefore, $r = 2$

$$Volume = \dfrac{4\pi r^3}{3}$$

$$Volume = \dfrac{4\pi (2)^3}{3}$$

$$Volume = \dfrac{4\pi (8)}{3}$$

$$Volume = \dfrac{32\pi}{3}$$, 4 times 8 is 32

$$Volume = \frac{32\pi}{3}$$

Volume ≈ 33.5 inches³

To determine the Volume of a Cylinder, we use the formula: $V = (A_{base})(height) \text{ or } V = \pi r^2 h$

You must be wondering why this is the formula. The reason this is the formula is because the Area of the base is multiplied by the height of a cylinder. The Base Area formula is determined by using the formula $Area_{Base} = \pi r^2$

Does this look familiar to you? It probably does, the base of a cylinder is a circle, the formula to find the area of a circle can be used to find the area of the base of a cylinder. Then you multiply it by the height to basically "Fill in" the shape from bottom to top.

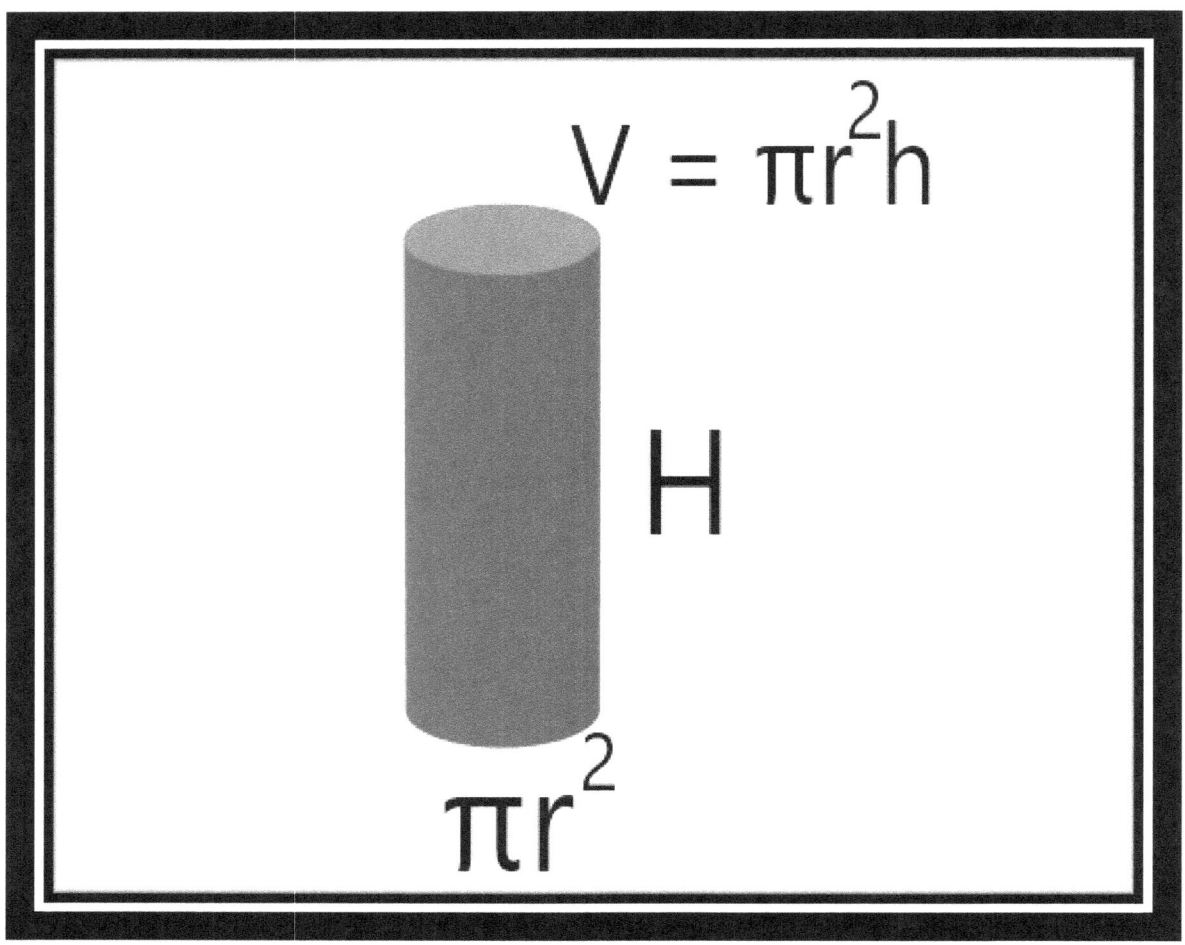

Example 1: Find the volume of a cylinder with a radius of 2cm and a height of 5cm.

- Step 1: Find radius and height. r = 2, h = 5
- Step 2: Plug "r" and "h" into the cylinder formula and solve for the volume.

$V = \pi(2)^2(5)$
$V = \pi(4)(5)$
$V = 20\pi$

$V \approx 62.8 \; cm^3$

Surface Area

Surface area is defined as the measure of the total area that the surface of the object occupies. Think about it this way, when you are wrapping Christmas gifts, surface area would be considered the total area of the "wrapping paper" that is covering your present. Volume would be the 3-dimensional space of the present. We will start with the Formula for the Surface area of the cube.

$SA = 6a^2$

For example, if the length of the cube is 3cm then, a=3.

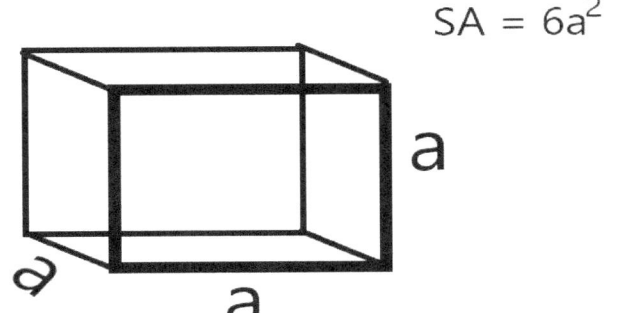

$SA = 6(3)^2$

$SA = 6(9)$

$SA = 54 cm^2$

To determine the Surface Area of a Sphere, we use the formula $A = 4\pi r^2$

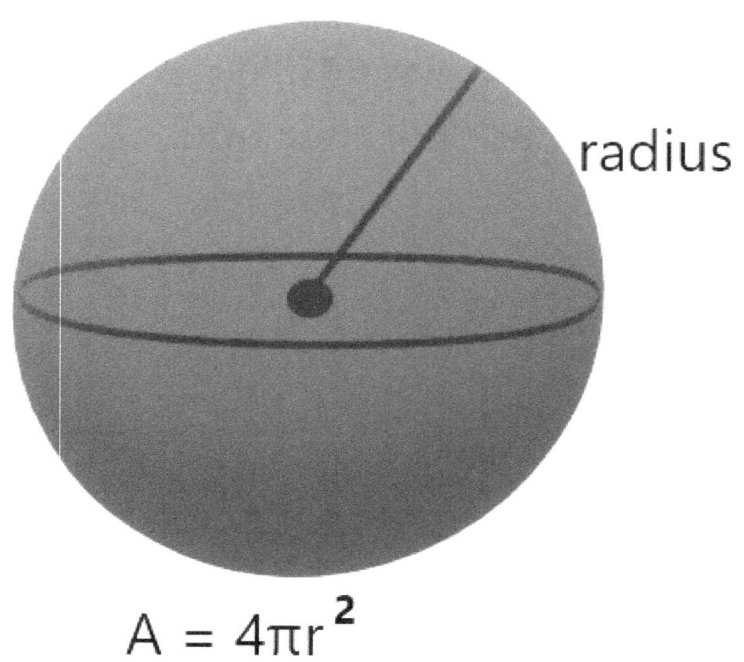

$A = 4\pi r^2$

Example 1: Find the Surface Area of a Sphere with a Diameter of 12 meters.

- Diameter is 12, we need the radius, so we divide the diameter by 2 to find it. $r = \dfrac{d}{2} = \dfrac{12}{2} = 6 \; meters$
- $SA = 4\pi r^2$

$$SA = 4\pi(6)^2$$
$$SA = 4\pi(36)$$
$$SA \approx 452.4 \; m^2$$

We will now look at determining the Surface Area of a Cylinder.

The formula is: $A_{total} = 2A_{base} + A_{lateral\;surface}$
$$= 2\pi r^2 + 2\pi rh$$

I will begin by explaining the formula, the first part is $2A_{base}$ the reason for the "2" is because there are two bases on a cylinder: the one at the top, and the one at the bottom. Lateral surface is basically the surface area of the cylindrical body of the cylinder, the surface excluding the bases.

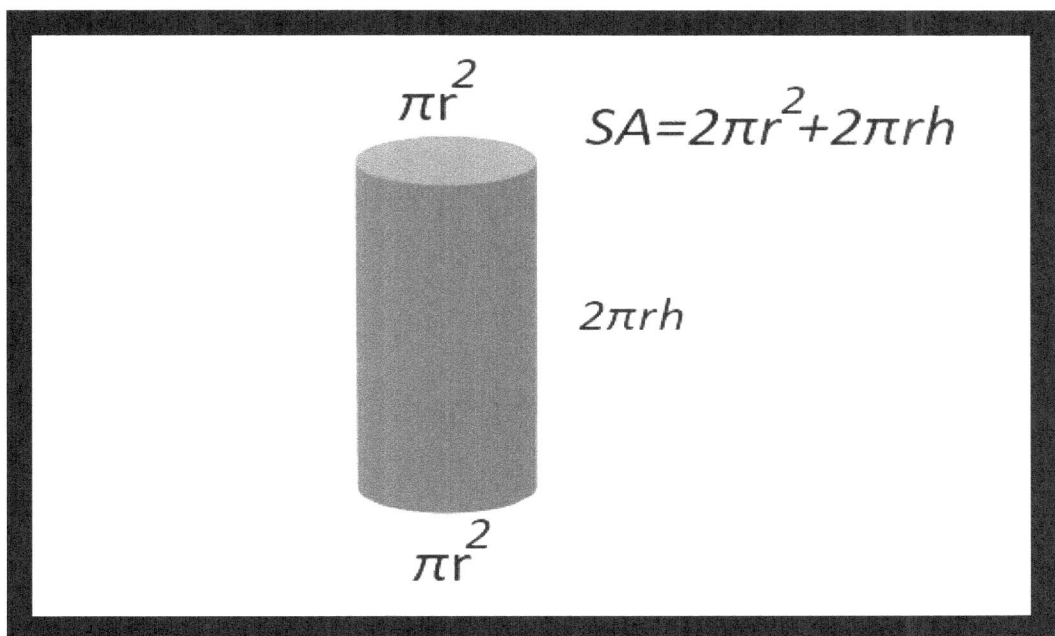

Example 1: Find the total Surface Area of a cylinder with a base of 5 cm, and a height of 7 cm.

r = 5, h = 7

$SA = 2\pi r^2 + 2\pi rh$
$SA = 2\pi(5)^2 + 2\pi(5)(7)$
$SA = 2\pi(25) + 2\pi(35)$
$SA = 50\pi + 70\pi$
$SA = 120\pi$
$SA \approx 376.99 \; cm$

Practice Questions

1. Determine the Volume of a perfect Cube with a width of 5cm.

2. Determine the Surface Area of a cube with a length of 4cm.

3. Determine the Volume of a Sphere with a diameter of 8m.

4. Determine the Surface Area of a Sphere with a radius of 4m.

5. Determine the Volume of a Cylinder with a base radius of 4cm and a height of 5cm.

6. Determine the Surface Area of a Cylinder with a base diameter of 8cm and a height of 5cm.

7. *Challenge Question:* Determine the expression for Volume of a Cylinder with a base radius of **x** cm, and a height of **3** cm.

RADICALS
Chapter 7

In this chapter we will be learning about radicals. Before we start learning how to simplify radical expressions, I think you should familiarize yourself with some vocabulary you will come across when learning about Radicals.

- Radicand: The expression under the radical sign ($\sqrt{}$).
- Index: The number that represents the numerical root of the radical expression. Ex: $\sqrt[4]{x}$ The 4 is the index, this means the 4th root of x.
- Rational Exponent: An exponent that is a fraction. You will learn in this unit how to convert rational exponents to a radical expression: $x^{\frac{2}{3}}$ 2/3 is the exponent.
- Radical Equation: An equation with a variable in the radicand.

    The picture on the next page shows all the components of a radical equation

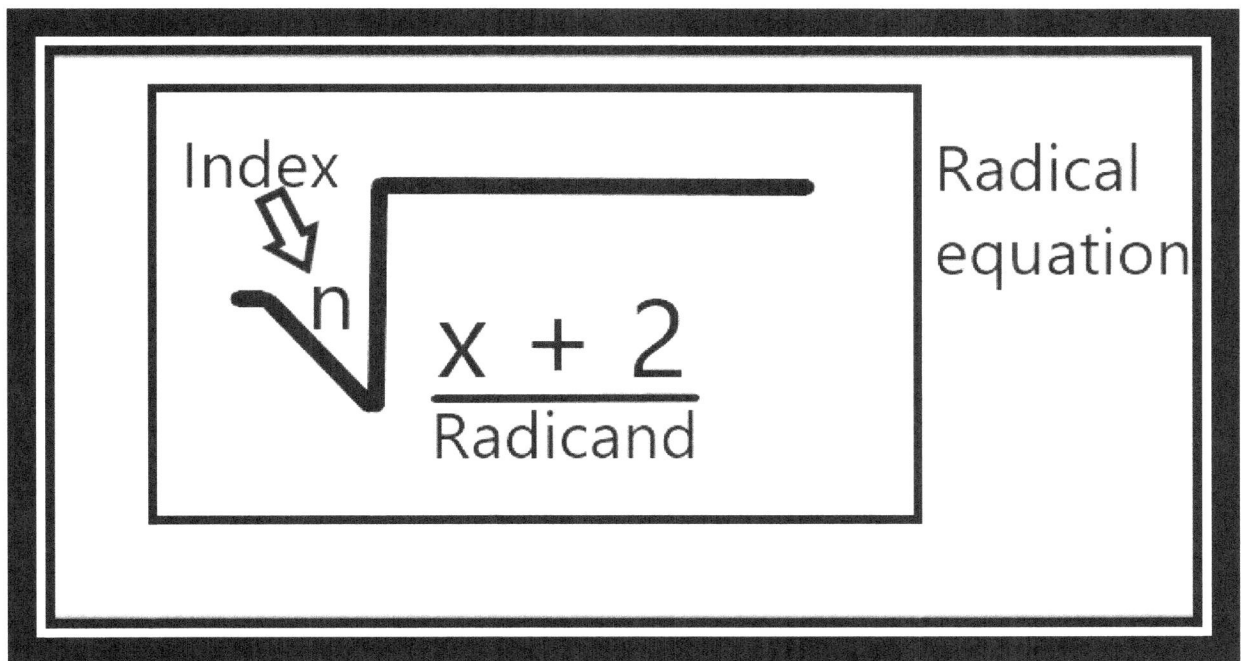

If there is no number in the index, then the index is assumed to be 2.

Now that we have looked at the basic vocabulary of Radicals, we will now learn how to simplify radicals.

Writing Mixed Radicals as Entire radicals

What is an **entire radical**? An entire radical is a radical with no coefficient in front. For example, $\sqrt{3}$ is an entire radical because there is no number in front. Here are some more examples of an entire radical:

- $\sqrt[3]{5}$ is an entire radical
- $\sqrt[4]{6}$ is an entire radical

- $2\sqrt{3}$ is not an entire radical because of the two in front. This is a mixed radical. A mixed radical is a radical with a number in front other than 1.

Sometimes you will be asked to convert mixed radicals to entire radicals, to do this you basically multiply the number in front of the radical by itself the number of times that is denoted in the index. Then you put that number inside the radicand and multiply that number by the other coefficient that is already inside of the radicand. For example, in $2\sqrt[3]{6}$ the index is 3, and the number in front of the radical is 2; therefore, you multiply 2 by itself 3 times, so, $(2)(2)(2) = 2^3 = 8$, then you put 8 inside the radical and multiply it by 6 like this, $\sqrt[3]{(8)(6)} = \sqrt[3]{48}$. Check on your calculator to ensure you got the right answer, you should get the same result on your calculator for both, the entire radical, and the mixed radical. Here are some more examples:

- $3\sqrt{5} = \sqrt{(3)(3)(5)} = \sqrt{45}$
  3 is multiplied by itself 2 times since the index is assumed to be 2.

- $2\sqrt[3]{4} = \sqrt[3]{(2)(2)(2)(4)} = \sqrt[3]{32}$

2 is multiplied by itself 3 times because the index of the radical is 3.

- $3\sqrt[4]{5} = \sqrt[4]{(3)(3)(3)(3)(5)} = \sqrt[4]{405}$

  3 is multiplied by itself 4 times since the index of the radical is 4.

Writing Entire Radicals as Mixed radicals

In the previous section you learned how to write a mixed radical as an entire radical. In this section, we will do the opposite. To do this, we must use the skill we learned in *Chapter 5* prime decomposition of integers.

For example, $\sqrt{20}$ we see that the number in the radicand is 20, and the index is 2. So, what we do now is we write all the factors of 20, and we look for factors that appear the number of times in the index, in this case, 2 times. The solution continues on the next page…

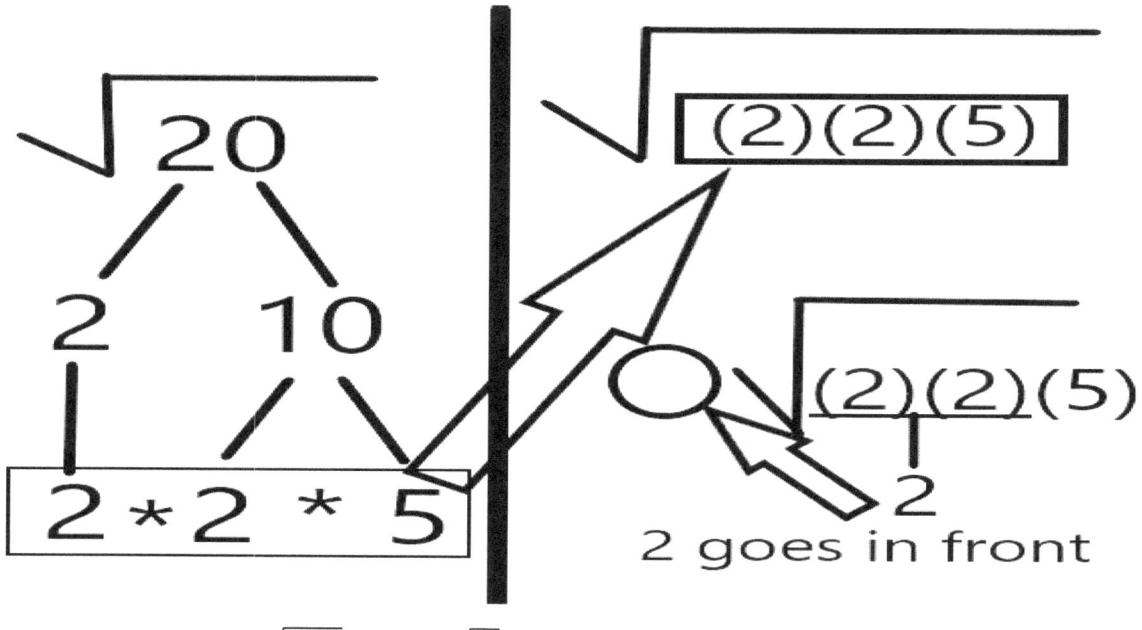

2 goes in front

Final answer $\sqrt{20} = 2\sqrt{5}$

As you can see, even though you found 2 as a factor twice the number in front was 2. The reason for this is because to remove the (2)(2) from the radicand you took the square root of this number and put the square root of 4 in front of the radical. $\sqrt{(2)(2)} = \sqrt{4} = 2$.

5 does not have a perfect square root; therefore, it stays inside the radicand.

Here are some more examples:

- $\sqrt[3]{54} = \sqrt[3]{(3)(3)(3)(2)}$ *3 appears 3 times,* this means that the 3 can go outside of the radical, and the 2 stays inside. This will give you, $3\sqrt[3]{2}$

- $\sqrt{8} = \sqrt{(2)(2)(2)}$ be very careful here, the index is 2. You are only looking for a factor that appears twice, and 2 appears 3 times. What you do is you basically only take 2 of the two's and you keep one of the two's inside the radicand.

$\sqrt{8} = \sqrt{(2)(2)(2)} = 2\sqrt{2}$

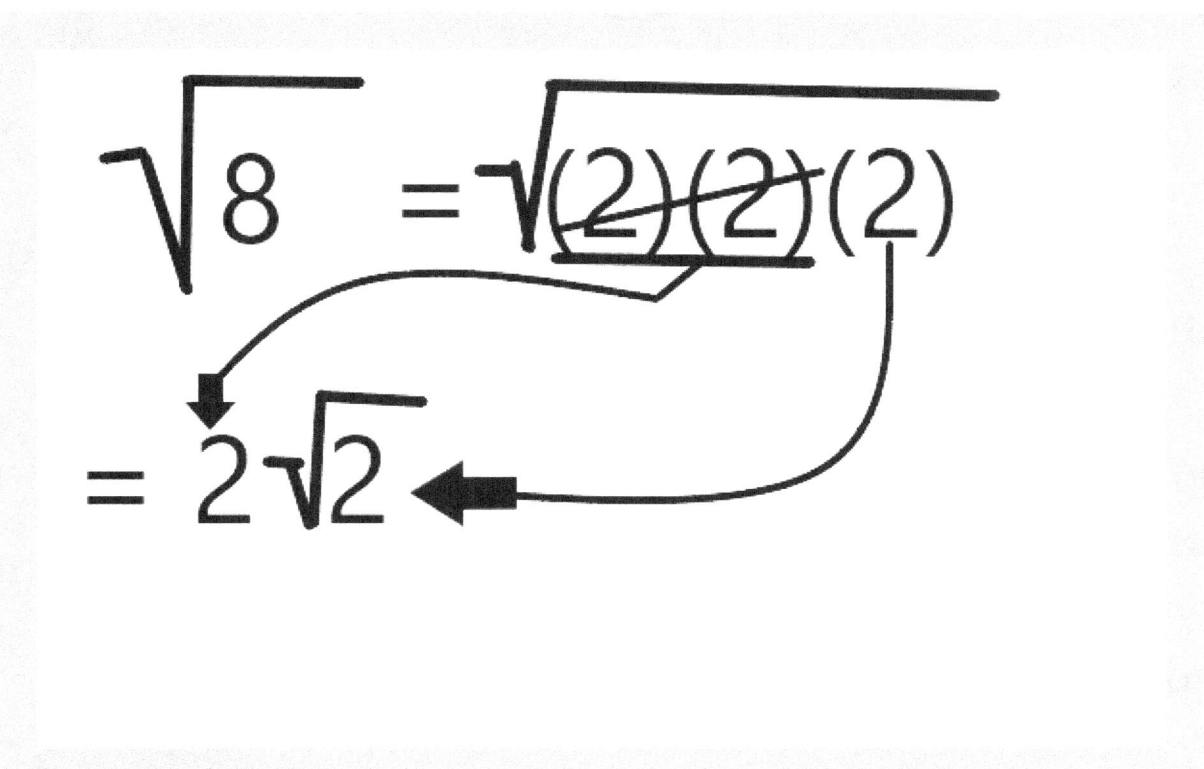

What about a variable expression?

$\sqrt{x^2 y} = \sqrt{(x)(x)(y)} = x\sqrt{y}$

$$\sqrt{x^2 y} = \sqrt{(x)(x)(y)} = x\sqrt{y}$$

Rational Exponents

As mentioned earlier in this chapter, rational exponents are exponents in the form of $x^{\frac{m}{n}}$, where $\frac{m}{n}$ is the exponent. To write this as a radical expression, you basically take the denominator of the exponent and make it the index, and you raise the radical to the power of the numerator. For example, $x^{\frac{m}{n}} = \sqrt[n]{x^m}$ or $(\sqrt[n]{x})^m$

Here are some more examples:

- $8^{\frac{2}{3}} = \sqrt[3]{8^2} = \sqrt[3]{(8)(8)} = \sqrt[3]{64} = 4$

  Or, $8^{\frac{2}{3}} = (\sqrt[3]{8})^2 = (\sqrt[3]{8})^2 = (2)^2 = 4$

- $4^{\frac{1}{2}} = \sqrt{4} = 2$
- $y^{\frac{5}{n}} = \sqrt[n]{y^5}$ or $(\sqrt[n]{y})^5$

Negative Exponents

We will learn how to simplify negative exponents in this section. Here is the general form, $x^{-n} = \dfrac{1}{x^n}$

For example, $2^{-1} = \dfrac{1}{2^1} = \dfrac{1}{2}$

Here are some more examples:

- $3^{-3} = \dfrac{1}{3^3} = \dfrac{1}{(3)(3)(3)} = \dfrac{1}{27}$
- $x^{-2} = \dfrac{1}{x^2}$
- $2^{-y} = \dfrac{1}{2^y}$

- $2^{-\frac{2}{3}} = \dfrac{1}{2^{\frac{2}{3}}} = \dfrac{1}{(\sqrt[3]{2})^2}$ or $\dfrac{1}{\sqrt[3]{2^2}} = \dfrac{1}{(\sqrt[3]{4})}$

Eventually you will learn how to remove a radical from the denominator, for now don't worry about it.
Try these two practice questions, the answers are on the next page,

1. x^{-3}
2. y^{-z}

The answer to question 1 is $\dfrac{1}{x^3}$

The answer to question 2 is $\dfrac{1}{y^z}$

Here is a picture summarizing the two new kinds of exponents you were introduced to,

Rational Exponent:

$$X^{\frac{m}{n}} = \sqrt[n]{X^m} = \left(\sqrt[n]{X}\right)^m$$

Negative Exponent:

$$X^{-n} = \frac{1}{X^n}$$

Practice Questions

1. Write the following mixed radicals as entire radicals:
   a) $3\sqrt{3}$
   b) $x\sqrt{x}$
   c) $5\sqrt{2}$
   d) $y\sqrt{x}$
   e) $2\sqrt[3]{3}$

2. Write the following entire radicals as mixed radicals:
   a) $\sqrt{28}$
   b) $\sqrt{z^3}$
   c) $\sqrt[3]{24}$
   d) $\sqrt{18}$

3. Write with a positive exponent:
   a) x^{-4}
   b) j^{-x}
   c) 3^{-3}
   d) 4^{-2}

4. Write the following rational exponents as a radical expression:
   a) $x^{\frac{y}{z}}$
   b) $3^{\frac{2}{3}}$

5. Challenge Questions:
   a) Write $x^{-\frac{3}{4}}$ as a radical expression without the negative exponent.
   b) Write this expression as a rational exponent.
   $(\sqrt[6]{x})^5$

Bonus Chapter: Trigonometry (11)

From: "High School Math Made Understandable Book 2: Math 11 & 12"

In grade 10, you learned a little bit about the Sine, Cosine and Tangent ratios. In this Chapter, you will be extending this knowledge even further. This is a very important chapter; therefore, the rate of progression in this chapter will be very gradual. We will be learning about angles in

Standard Position (when the initial arm lies on the positive x-axis). First, we will look at how the four quadrants on the cartesian plane are divided.

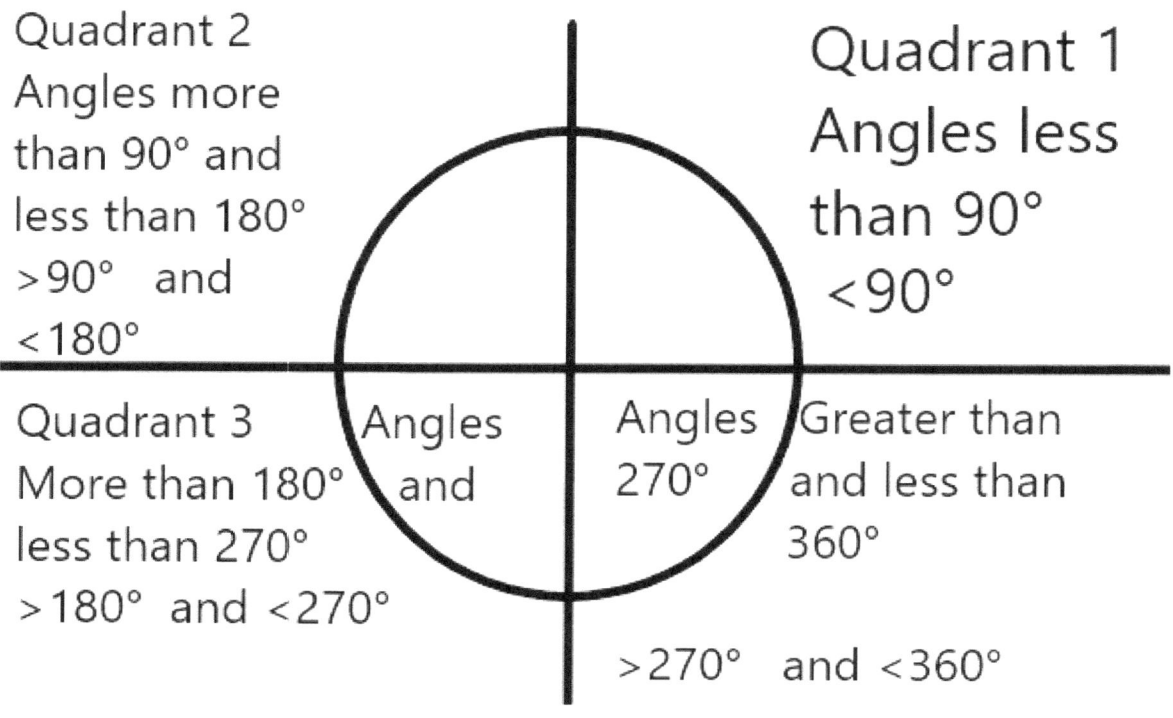

As you can see, there is a circle that is divided into four sections (the four quadrants), and each quadrant has a range of 90°. 90° lies on the positive y-axis, 180° lies on the negative x-axis, 270° lies on the negative y-axis, and 0° and 360° lies on the positive x-axis. When we sketch an angle in Standard position, something important to note is that when your angle is positive the rotation is counter clockwise, and when the angle is negative the rotation is clockwise.

θ (This symbol is called "Theta"), this fancy symbol simply means "Angle."

Here are two examples of drawing angles in standard position.

Example 1: $\theta = 45°$

This means that the angle we are drawing is 45° in Standard position. On the next page you will be given the step-by-step process.

- Step 1: Draw the 4 quadrants of the Cartesian Plane,

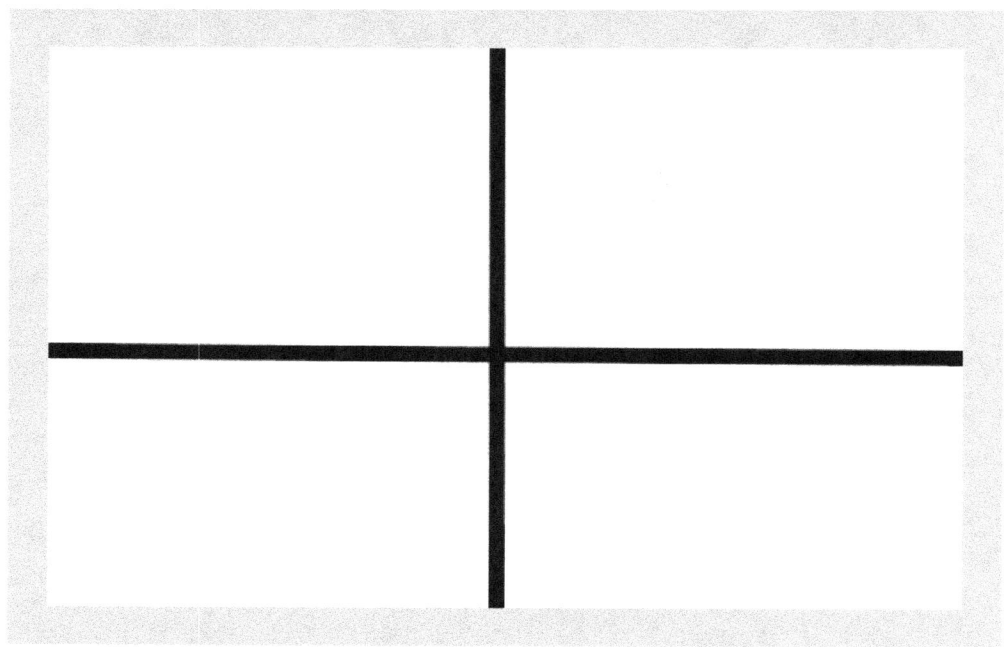

- Step 2: Determine which quadrant θ lies in. θ in this case is 45°, therefore, θ lies in Quadrant 1.

- Step 3: Determine where θ lies in Quadrant 1, 45° is half of 90°. With this being said, 45° is half-way to 90°. Draw the terminal arm halfway to 90° diagonally.

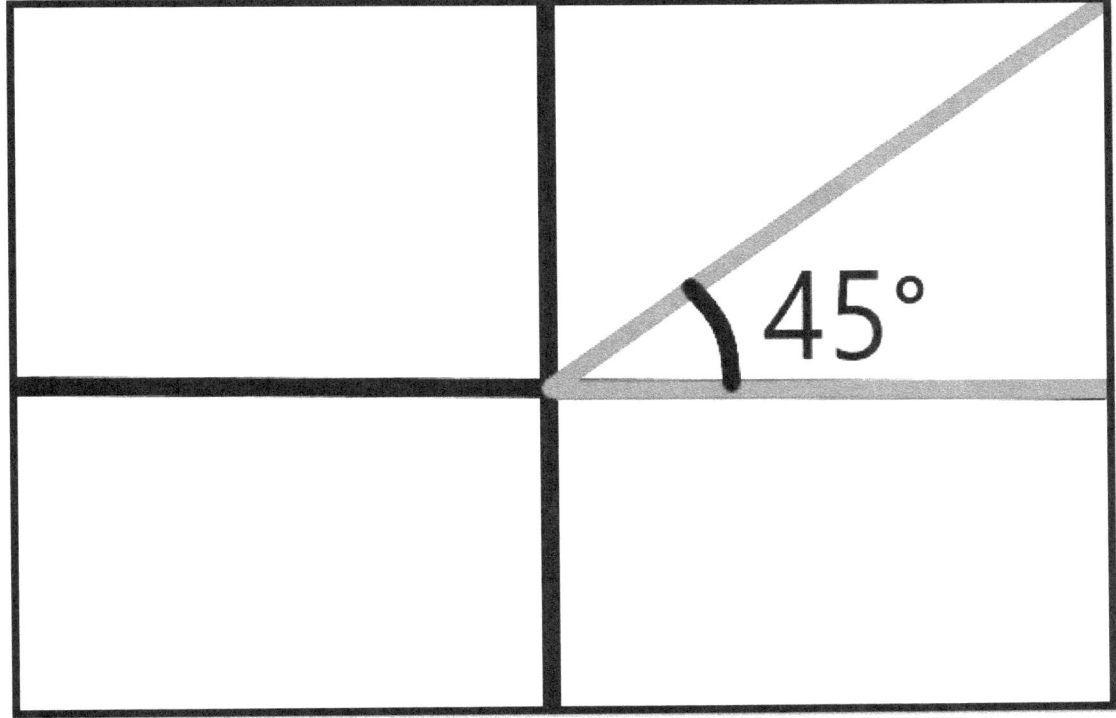

Notice how it is the Terminal arm that determines the angle of the sketch.

Example 2: $\theta = 135°$

- Step 1: Determine which quadrant 135° lies in. It is located in Quadrant 2 since any angle greater than 90° and less than 180° is found in that quadrant.

Step 3: Determine where θ lies in Quadrant 2, 135° is halfway from 90° to 180°. Draw the terminal arm diagonally in-between the 90° and 180°.

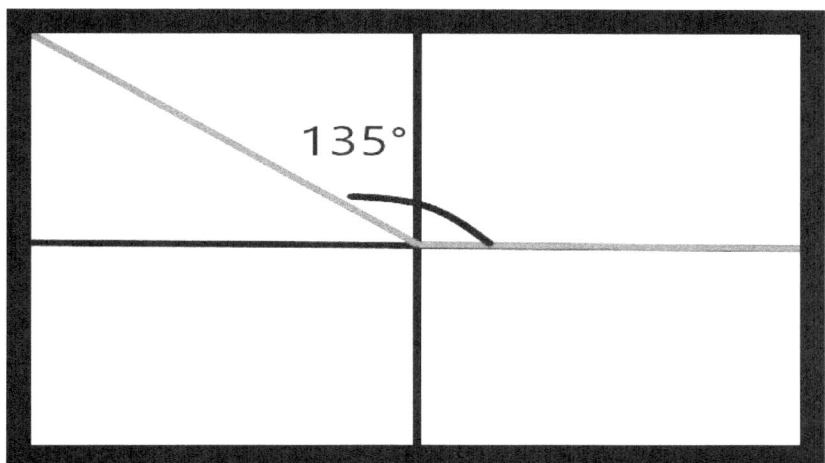

Reference Angles

A reference angle is the angle that the terminal arm makes with the x-axis. Here are two examples of reference angles. Let a represent the reference angle.

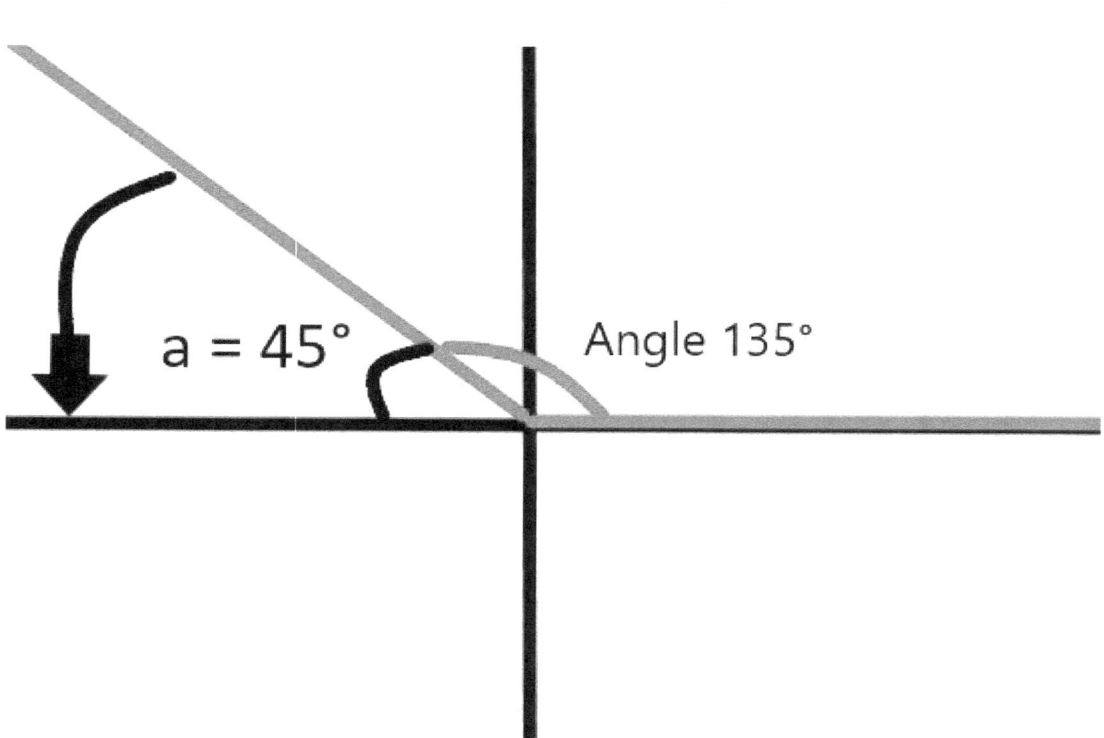

If the angle drawn is less than 180° then you basically subtract the angle drawn from 180° giving you the reference angle. Ex: $\theta = 135°$

$a = 180° - \theta = 180° - 135° = 45°$

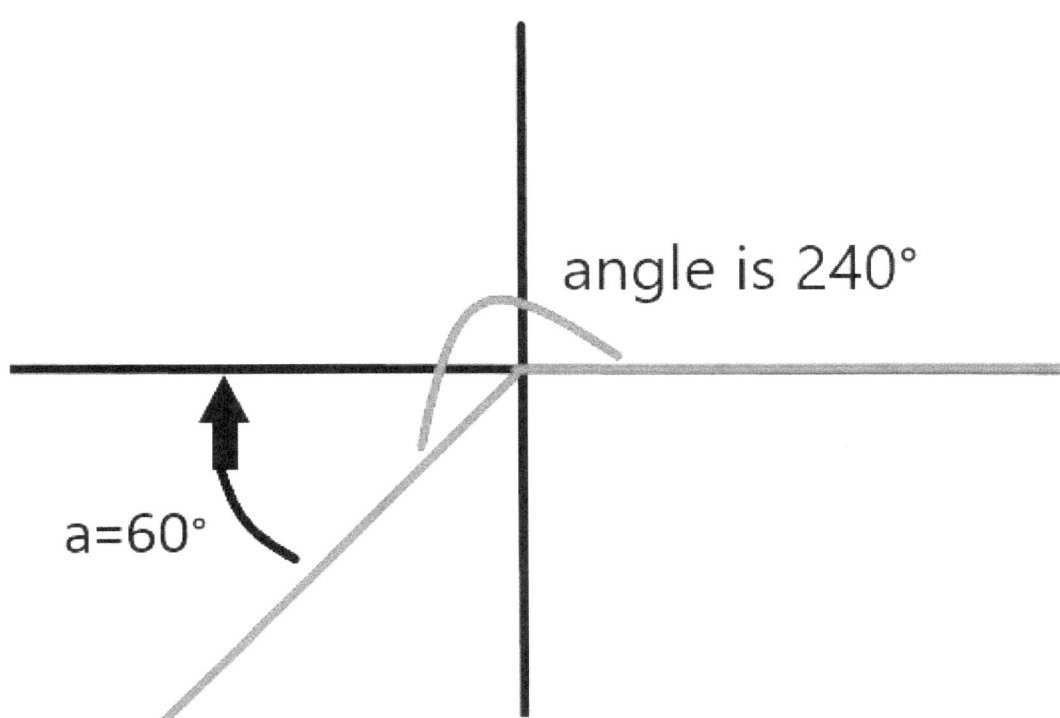

If the angle drawn is more than 180° and less than 270° then you basically subtract 180° from the angle drawn, giving you the reference angle. Ex: $\theta = 240°$

$a = \theta - 180° = 240° - 180° = 60°$

When the angle drawn is located in Quadrant 1, then the reference angle is equal to the angle of the terminal arm in Quadrant 1. For instance, if the angle drawn is 30° then the reference angle is 30°.

Sine, Cosine and Tangent Ratios

In this section we will be using angles in standard position to determine the exact value of Trigonometric ratios. Exact values mean no decimals, the exact precise value. This chapter as I have previously mentioned is very important, you will see these angles in standard position and Trigonometric ratios come back to haunt you in Grade 12.

$$\sin\theta = \frac{opp}{hyp} = \frac{y}{r}$$

$$\tan\theta = \frac{opp}{adj} = \frac{y}{x}$$

$$\cos\theta = \frac{Adj}{hyp} = \frac{x}{r}$$

$$r = \sqrt{x^2 + y^2}$$

y-coordinate is the rise, x-coordinate is the run, and r is the hypotenuse(diagonal). You can determine the r value by using the Pythagorean Theorem.

Example 1: If the coordinate is (3,4),
$r = \sqrt{x^2 + y^2} = \sqrt{3^2 + 4^2} = \sqrt{9 + 16} = \sqrt{25} = 5$

$r = 5$ hypotenuse is 5

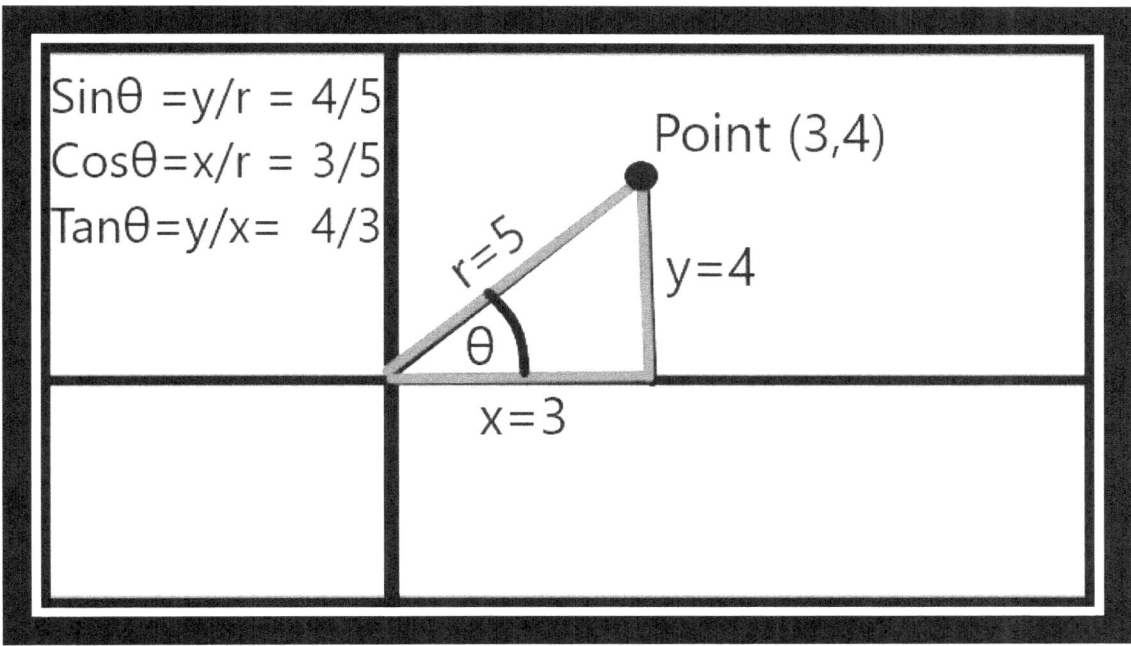

Example 2: Determine the exact value of $\sin\theta, \cos\theta, \tan\theta$, if the terminal arm of an angle in standard position passes through the Point (4,5).

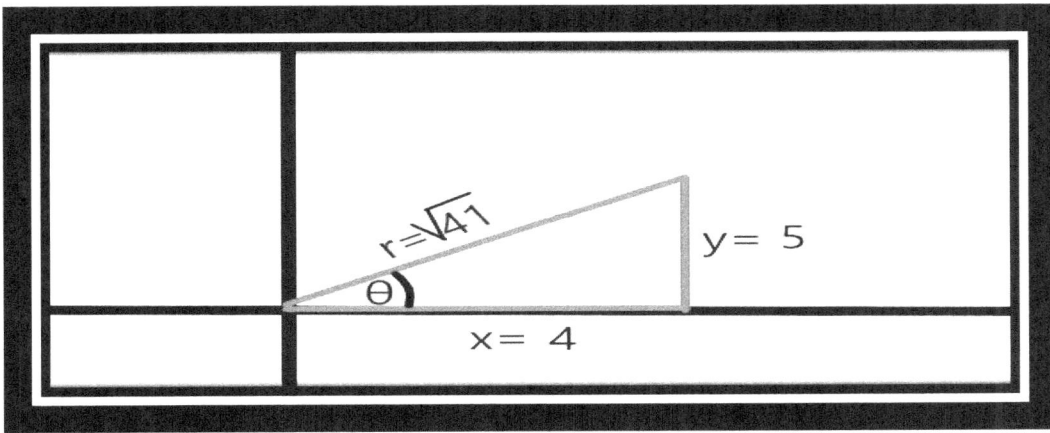

$$r = \sqrt{x^2 + y^2} = \sqrt{16 + 25} = \sqrt{41}$$
$$r = \sqrt{41}$$

Keep the "r" value as a radical, and plug the values into the corresponding ratios:

$$sin\theta = \frac{y}{r} = \frac{5}{\sqrt{41}}$$

$$cos\theta = \frac{x}{r} = \frac{4}{\sqrt{41}}$$

$$tan\theta = \frac{y}{x} = \frac{5}{4}$$

Sine Law

In this section we will learn about the Sine Law, this law is useful for finding a side or an angle of a triangle that is not a right-triangle (triangle in which one of its angles is 90°).

The Sine Law states that $\frac{a}{sinA} = \frac{b}{sinB} = \frac{c}{sinC}$

You will see a few examples that will help you understand what this means.

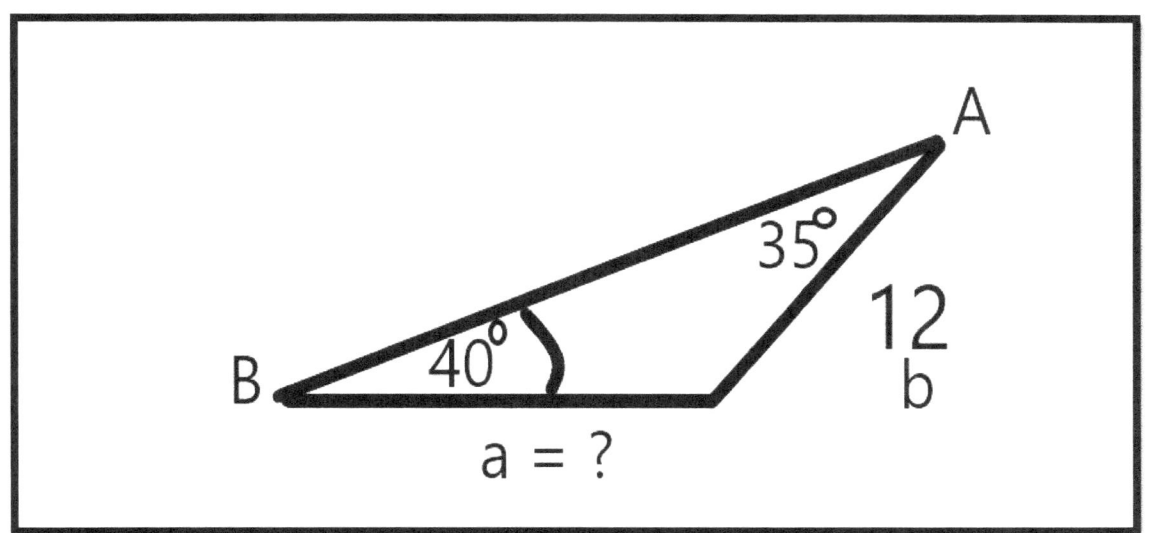

As you can see, the question is asking you to find the value of "a"

Sin B = 40° $Sin\ A = 35°$ b = 12   a = ?

$$\frac{a}{sinA} = \frac{b}{sinB}$$

$$\frac{a}{sin35°} = \frac{12}{sin40°}$$

We now rearrange the formula to solve for "a", we multiply both sides by sin35°,

$$(sin35)\frac{a}{sin35°} = \frac{12}{sin40}(sin35)$$

$$a = \frac{12sin35}{sin40}$$

$$a = 10.7$$

The length of side "a" is 10.7, that's it we have found side "a"! A good way to verify your answer is by substituting "a" inside of the formula and you should get the same ratio for both sides of the equation (should be very close).

Example 2: $$\frac{Sin38}{50} = \frac{SinB}{35}$$

We are looking for angle B, therefore, we will rearrange the formula to solve for Sin B, and then once you have the ratio use the Inverse Sin function(Sin^{-1}) on your calculator to solve for "Angle B"

$$\frac{sin38}{50}(35) = \frac{SinB}{35}(35)$$

$$\frac{35 sin38}{50} = SinB$$

$$SinB = 0.43096032...$$

$$B = Sin^{-1}(0.43096032...)$$

$$B \approx 25.53° \approx 26°$$

Cosine Law

The *Law of Cosines* says that:

$$c^2 = a^2 + b^2 - 2ab\cos(C)$$

This law helps us solve triangles that do not have an angle of $90°$.

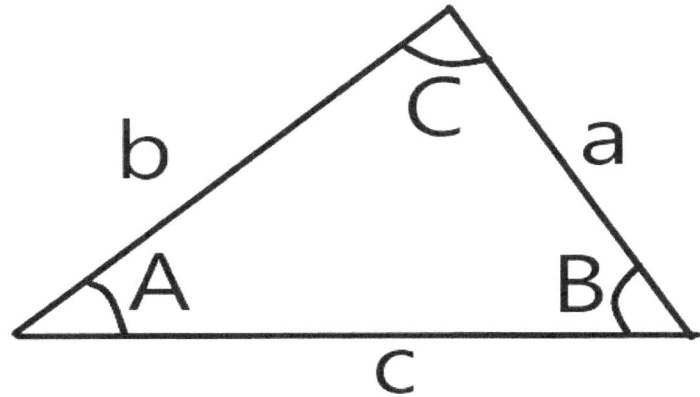

We will look at two examples of how this law can be applied:

Example 1: Find the length of side "c"

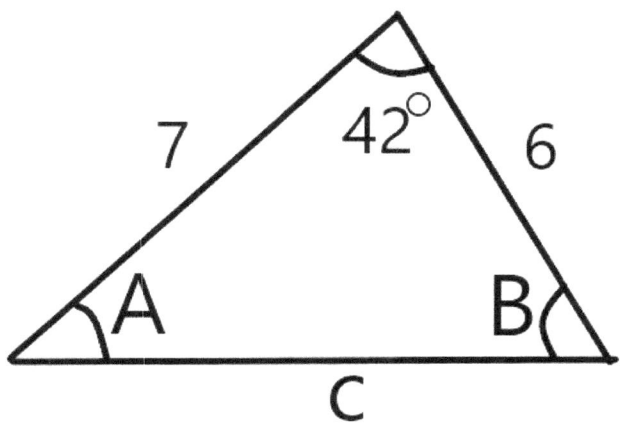

We will use the formula: $c^2 = a^2 + b^2 - 2abCos(C)$, since we are trying to find side "c."

Angle C $= 42°$ $\quad Cos(C) = Cos(42°)$

$a = 7 \quad b = 6 \quad$ plugging this into our formula we get,

$c^2 = (7)^2 + (6)^2 - 2(7)(6)Cos(42°)$

$c^2 = 49 + 36 - (84)(\cos(42°))$

$c^2 = 85 - 62.42 \quad c^2 = 22.58$

$c = \sqrt{22.58} \quad\quad c \approx 4.75$

There are some other rearrangements of this formula:

$a^2 = b^2 + c^2 - 2bcCos(A)$

$b^2 = a^2 + c^2 - 2acCos(B)$

Basically, the angle you use in the formula is the angle opposite to the side you are trying to find the length of. This formula can even be rearranged in order to find the measure of the angle opposite to a given side. Here are the formulas:

$$Cos(C) = \frac{a^2 + b^2 - c^2}{2ab} \qquad Cos(B) = \frac{c^2 + a^2 - b^2}{2ca}$$

$$Cos(A) = \frac{b^2 + c^2 - a^2}{2bc}$$

Example 2: Find Angle C.

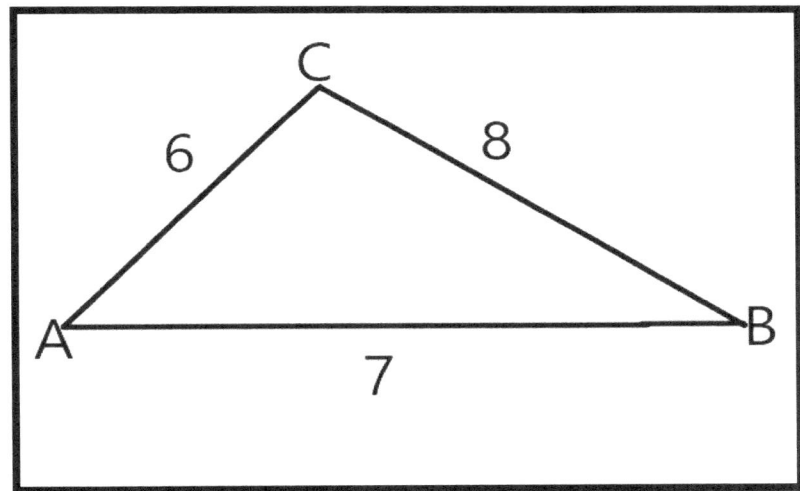

$a = 8 \quad b = 6 \quad c = 7$

Which formula should we use? We should use

$$Cos(C) = \frac{a^2 + b^2 - c^2}{2ab}$$, because we are trying to find Angle C.

Let's plug $a, b,$ and c into the formula:

$$Cos(C) = \frac{(8)^2 + (6)^2 - (7^2)}{2(8)(6)}$$

$$Cos(C) = \frac{64 + 36 - 49}{96} = \frac{100 - 49}{96} = \frac{51}{96}$$

$$Cos(C) = \frac{51}{96}$$ Use the inverse cosine function

$$C = Cos^{-1}\left(\frac{51}{96}\right) \quad C \approx 57.9°$$

Practice Questions

1. Draw an angle of 60° in standard position.

2. Draw an angle of 30° in standard position.

3. Draw an angle of 150° in standard position, what is the reference angle?

4. Determine the exact values of $sin\theta, cos\theta, tan\theta$ if the terminal arm of an angle in standard position passes through point P$(1, \sqrt{3})$. Hint: $\theta = 60°$,

5. Find the length of b.

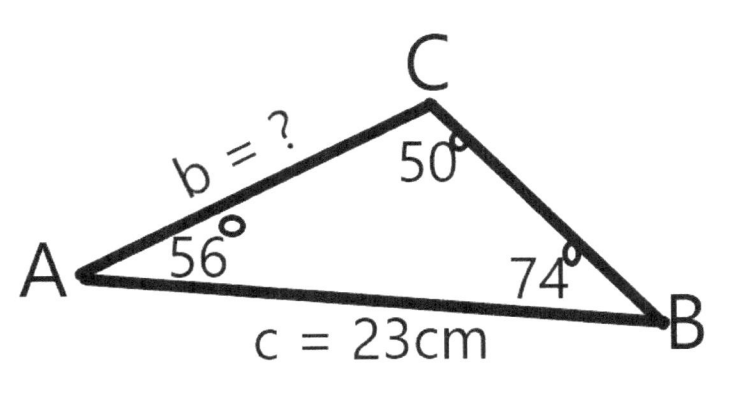

6. $\dfrac{18}{Sin36} = \dfrac{a}{Sin120}$

Find a.

7. What is the reference angle of 120°?

CONCLUSION

You have finished this book, congratulations! This is the first book in the *High School Math Made* Understandable series, I hope it helped you through your first two years of math. If you liked this book, then I strongly recommend you buy *High School Math Made Understandable Book 2: Math 1 & 12*. This book is already available on Amazon , so if you are interested you can order one today!

Answers & Solutions to Practice Questions

Chapter 1: POLYNOMIALS (Introduction):

1. $2x + 2x + 3x + y = 7x + y$
   Collect like terms:
   $2x + 2x + 3x + y = 7x + y$ since $2 + 2 + 3 = 7$

2. A) False   B) True   C) False

3. a) $3y - y = 3y - 1y = 2y$
   b) $2x^2 + 3x^2 = 5x^2$
   c) $ab + ab = 1ab + 1ab = 2ab$

4. a) $2x + 3y = 2(2) + 3(4) = 4 + 12 = 16$
   b) $y - x = (4) - (2) = 4 - 2 = 2$
   c) $x^2 = (2)^2 = (2)(2) = 4$
   d) $y^2 = (4)^2 = (4)(4) = 16$

5. a) $3x = 4 - x$	5. b) $2x + 2 = 8$
$3x + x = 4 - x + x$	$2x + 2 - 2 = 8 - 2$
$4x = 4$	$2x = 6$
$\dfrac{4x}{4} = \dfrac{4}{4}$	$\dfrac{2x}{2} = \dfrac{6}{2}$

$x = 1$ $\qquad$ $x = 3$

Chapter 2: EXPONENT LAWS:

1. $y(xy^2) = (xy^2)(y^1) = xy^{2+1} = xy^3$
2. $\dfrac{a^3 c^2}{a^2} = a^{3-2} c^2 = ac^2$
3. $(y^2)^a = y^{2*a} = y^{2a}$
4. $\dfrac{x^2 y^2}{x^2 y^2} = x^{2-2} y^{2-2} = x^0 y^0 = 1$
5. $\dfrac{x^y}{x^z} = x^{y-z}$
6. $\dfrac{7^5}{7^2} = 7^{5-2} = 7^3 = (7)(7)(7) = 343$

Chapter 3: LINEAR FUNCTIONS:

1. a) 3   b) 1   c) 2   d) 4

2.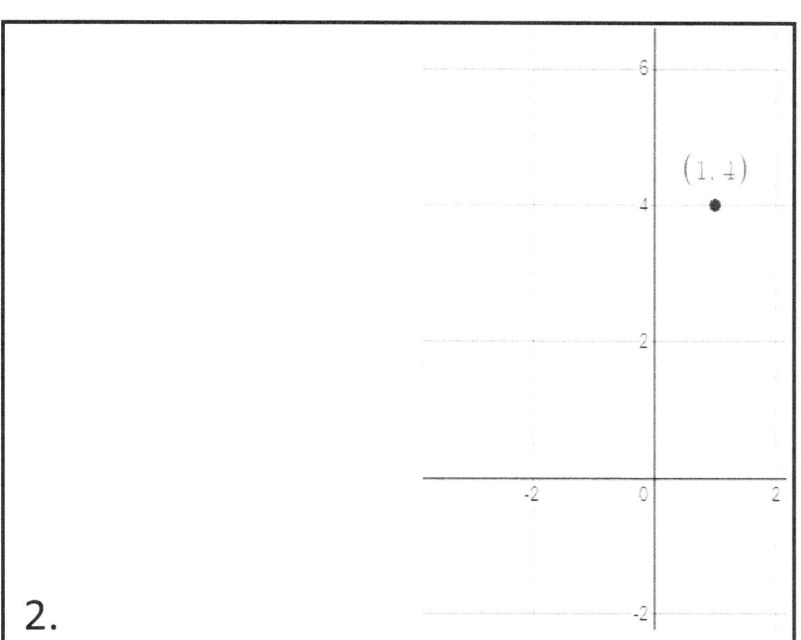

3. $y = 3x + 7$, because slope is 3, and y – intercept is 7.

4. $m = \dfrac{y_2 - y_1}{x_2 - x_1} = \dfrac{6 - 4}{7 - 3} = \dfrac{2}{4} = \dfrac{1}{2}$, m=1/2

5. $y - 7 = x$ $y - 7 = 0$

   $y - 7 + 7 = 0 + 7$ $y = 7$

   y-intercept is 7.

6. $y = x - 3$ $0 = x - 3$
   $0 + 3 = x - 3 + 3$ $x = 3$
   x-intercept is 3.

7. $f(2) = 2(2) + 2$

$$f(2) = 4 + 2 = 6$$
$$f(2) = 6 \qquad (2,6)$$

1. Domain of $f(x) = 2x + 2$, $\{x|xER\}$
   Range of $f(x) = 2x + 2$, $\{x|xER\}$

2. Domain: $\{1,3,5,7\}$
   Range: $\{2,4,6,8\}$

Chapter 4: Trigonometry:

1. $sinx = \dfrac{opp}{hyp} = \dfrac{6}{10} = 0.6 \qquad cosx = \dfrac{adj}{hyp} = \dfrac{8}{10} = 0.8$

   $tanx = \dfrac{opp}{adj} = \dfrac{6}{8} = 0.75 \qquad x = 37°$

2. $\tan(A) = \dfrac{7}{6}$ $A = \tan^{-1}\left(\dfrac{7}{6}\right)$

   $A \approx 49.4°$

3. $\sin(45°) = \dfrac{6}{x}$ multiply both sides by x

   $x * \sin(45°) = \dfrac{6}{x} * x$

   $\dfrac{x * sin45°}{sin45°} = \dfrac{6}{\sin 45°}$ divide both sides by $sin45°$

   $x = \dfrac{6}{\sin 45°}$ $x \approx 8.49$

4. Solution may vary

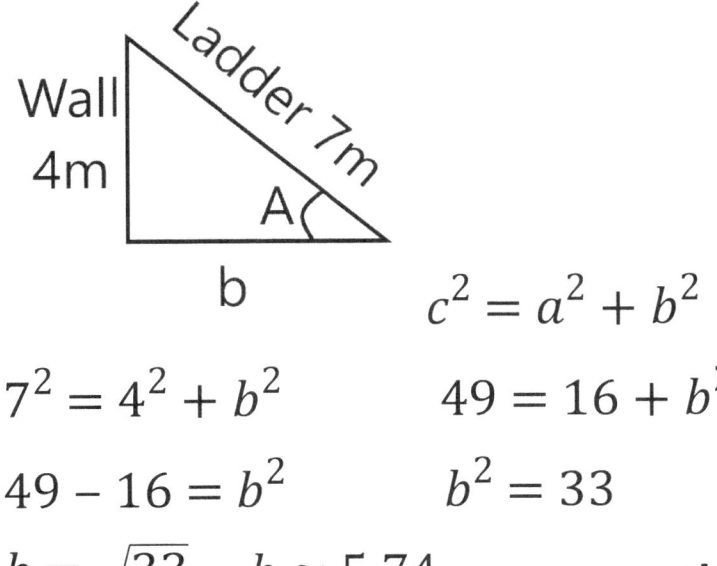

   $c^2 = a^2 + b^2$

   $7^2 = 4^2 + b^2$ $49 = 16 + b^2$

   $49 - 16 = b^2$ $b^2 = 33$

   $b = \sqrt{33}$ $b \approx 5.74$ meters   the base of the ladder is **5.74 meters from the wall.**

$$SinA = \frac{opp}{hyp} = \frac{wall}{ladder} = \frac{4}{7}$$

$$A = \sin^{-1}\left(\frac{4}{7}\right) \quad A \approx 35°$$

5.

a) $$\tan 26° = \frac{y}{2}$$

$$2 * \tan 26° = \frac{y}{2} * 2$$

$$2\tan 26° = y \quad y \approx 0.98$$

b) $$\cos(y) = \frac{1}{2}$$

$$y = \cos^{-1}\left(\frac{1}{2}\right) \quad y = 60°$$

c) $$\sin 35° = \frac{6}{y}$$

$$y * \sin(35°) = \frac{6}{y} * y \quad y * \sin(35°) = 6$$

$$y = \frac{6}{\sin(35°)} \quad y \approx 10.46$$

6. (Solution may vary)

$Angle\ B = 180° - 90° - 35° = 55°$

$a = \sin 35° = \dfrac{a}{9}$

$(9)(sin35°) = a$

$a = 5.16$

$b = cos35° = \dfrac{b}{9}$

$(9)(cos35°) = b$

$b = 7.37$

Chapter 5: Factoring Polynomials:

1. a) 60 = (30)(2) = (15)(2)(2) = (3)(5)(2)(2)

    b) 88 = (44)(2) = (22)(2)(2) = (11)(2)(2)(2)

    c) 390 = (195)(2) = (39)(5)(2) = (13)(3)(5)(2)

2. 48 = 3*2*2*2*2   GCF = 3*2*2*2 = 24

    72 = 3*2*2*2*3

3. a) $2x^2 + 4x = 2x(x+2)$
b) $3x + 3 = 3(x+1)$
c) $5x^2 - 15x = 5x(x-3)$
d) $y^2 + y = y(y+1)$
e) $2ab + b = b(2a+1)$

4. A) $x^2 + 11x + 30 = (x+6)(x+5)$

Two numbers that multiply to 30 and add up to 11. (6 & 5)

b) $y^2 + 17y + 60 = (y+5)(y+12)$
Two numbers that multiply to 60 and add up to 17 (5 & 12)

c) $x^2 - x - 6 = (x-3)(x+2)$
Two numbers that multiply to -6 and add up to -1. (-3 & 2)

5.a) $5x^2 - 7x + 2$

Two numbers that multiply to 10(since 5*2=10) and adds up to -7. (-5 and -2)

$[5x^2 - 5x] - 2x + 2$
$= 5x(x-1) - 2(x-1)$
$= (5x-2)(x-1)$

5.b) $6x^2 + 11x - 10$

6*-10= -60

Two numbers that multiply to -60 and add up to 11. (15 & -4)

$6x^2 - 4x + 15x - 10$

$= 2x(3x - 2) + 5(3x - 2)$

$= (2x + 5)(3x - 2)$

Chapter 6: Volume and Surface Area of Geometric Shapes:

1. $v = lwh$
   $v = (5)(5)(5)$
   $v = 125 cm^3$

2. $SA = 6a^2 = 6(4)^2$
   $6(16) = 96 cm^2$

3. $Diameter = 8m, \quad radius = \dfrac{diameter}{2} = \dfrac{8}{2} = 4m$

   $Volume = \dfrac{4\pi r^3}{3} = \dfrac{4\pi(4)^3}{3}$

   $= \dfrac{4\pi(64)}{3} = \dfrac{256\pi}{3}$

   $Volume \approx 268.08 m^3$

4. $r = 4m$
   $SA = 4\pi r^2$

$$SA = 4\pi(4)^2$$
$$SA = 4\pi(16)$$
$$SA = 64\pi \approx 201.06 m^2$$

5. $r = 4cm, \ h = 5cm$
$$V = \pi r^2 h$$
$$V = \pi(4)^2(5) = \pi(16)(5)$$
$$V = 80\pi$$
$$V \approx 251.33 \ cm^3$$

6. $r = \dfrac{d}{2} = \dfrac{8}{2} = 4, \quad h = 5cm$
$$SA = 2\pi r^2 + 2\pi rh$$
$$SA = 2\pi(4)^2 + 2\pi(4)(5)$$
$$SA = 2\pi(16) + 2\pi(20)$$
$$SA = 32\pi + 40\pi$$
$$SA = 72\pi$$
$$SA \approx 226.19 cm^2$$

7. $r = x, \quad h = 3cm$
$$V = \pi r^2 h$$
$$V = \pi(x)^2(3)$$
$$V = 3\pi x^2$$

Chapter 7: Radicals:

1. A) $3\sqrt{3} = \sqrt{3*3*3} = \sqrt{27}$
   b) $x\sqrt{x} = \sqrt{x*x*x} = \sqrt{x^3}$
   c) $5\sqrt{2} = \sqrt{5*5*2} = \sqrt{50}$
   d) $y\sqrt{x} = \sqrt{y*y*x} = \sqrt{y^2 x}$
   e) $2\sqrt[3]{3} = \sqrt[3]{2*2*2*3} = \sqrt[3]{24}$

2. A) $\sqrt{28} = \sqrt{2*2*7} = 2\sqrt{7}$
   b) $\sqrt{z^3} = \sqrt{z*z*z} = z\sqrt{z}$
   c) $\sqrt[3]{24} = \sqrt[3]{2*2*2*3} = 2\sqrt[3]{3}$
   d) $\sqrt{18} = \sqrt{3*3*2} = 3\sqrt{2}$

3. A) $x^{-4} = \dfrac{1}{x^4}$
   b) $j^{-x} = \dfrac{1}{j^x}$
   c) $3^{-3} = \dfrac{1}{3^3} = \dfrac{1}{(3)(3)(3)} = \dfrac{1}{27}$
   d) $4^{-2} = \dfrac{1}{4^2} = \dfrac{1}{(4)(4)} = \dfrac{1}{16}$

4. A) $x^{\frac{y}{z}} = \sqrt[z]{x^y}$ or $(\sqrt[z]{x})^y$
   b) $3^{\frac{2}{3}} = \sqrt[3]{3^2} = \sqrt[3]{9}$ or $(\sqrt[3]{3})^2$

5.a) $\quad x^{-\frac{3}{4}} = \frac{1}{x^{\frac{3}{4}}} = \frac{1}{\sqrt[4]{x^3}} \quad$ or $\quad \frac{1}{(\sqrt[4]{x})^3}$

b) $\quad (\sqrt[6]{x})^5 = x^{\frac{5}{6}}$

Bonus Chapter: Trigonometry 11

1.

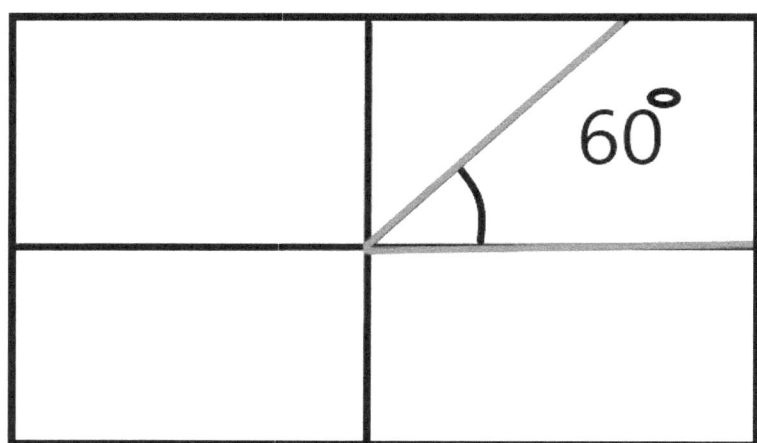

2.

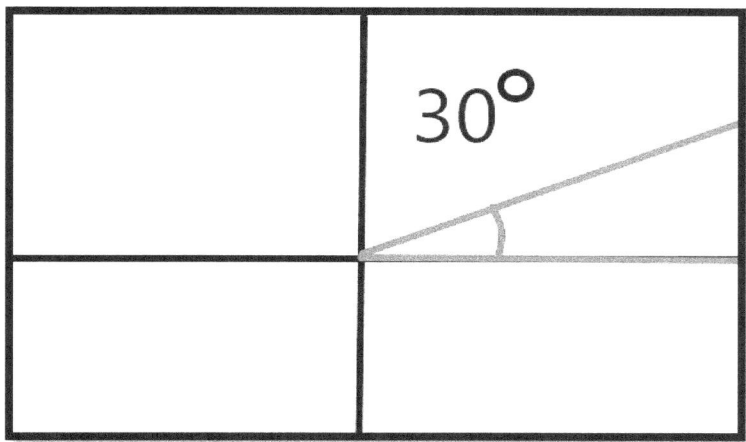

3.

4.

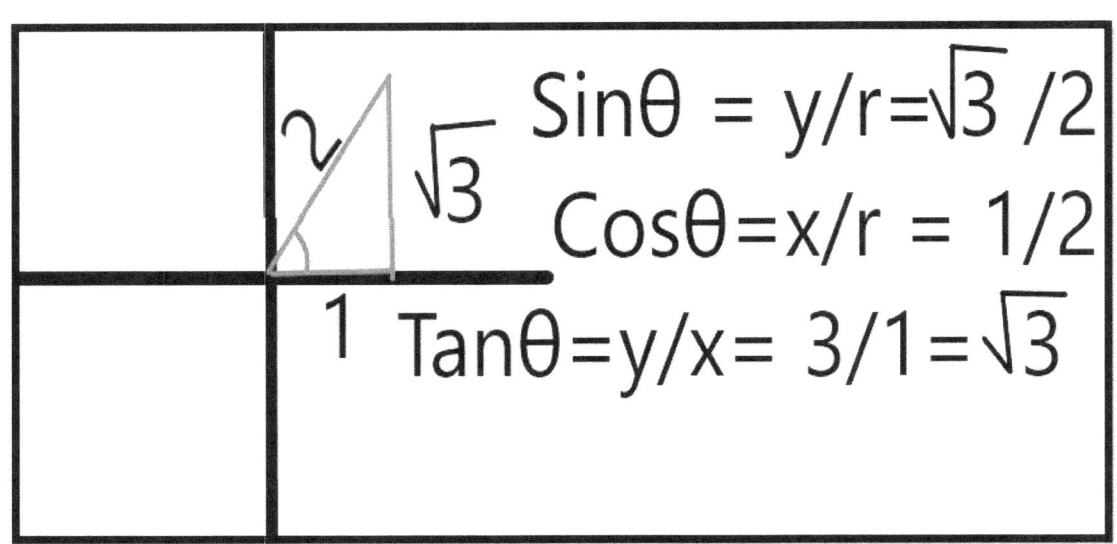

5. $\dfrac{23}{sin50} = \dfrac{b}{sin74}$

$\dfrac{23 sin74}{sin50} = b$

$b \approx 28.9\ cm$

6. $\dfrac{18}{sin36} = \dfrac{a}{sin120}$

$\dfrac{18 sin120}{sin36} = a$

$a \approx 26.5$

7. $Reference\ angle = 180° - 120° = 60°$

www.ingramcontent.com/pod-product-compliance
Lightning Source LLC
Chambersburg PA
CBHW060417220526
45465CB00008B/2925